Computational Methods Using MATLAB®

An introduction for physicists

Computational Methods Using MATLAB®

An introduction for physicists

P K Thiruvikraman

Department of Physics, BITS Pilani Hyderabad Campus, Telangana 500078, India

IOP Publishing, Bristol, UK

Multimedia content is available for this book from https://iopscience.iop.org/book/978-0-7503-3791-5.

ISBN   978-0-7503-3791-5 (ebook)
ISBN   978-0-7503-3789-2 (print)
ISBN   978-0-7503-3792-2 (myPrint)
ISBN   978-0-7503-3790-8 (mobi)

DOI   10.1088/978-0-7503-3791-5

Version: 20220301

IOP ebooks

British Library Cataloguing-in-Publication Data: A catalogue record for this book is available from the British Library.

Published by IOP Publishing, wholly owned by The Institute of Physics, London

IOP Publishing, Temple Circus, Temple Way, Bristol, BS1 6HG, UK

US Office: IOP Publishing, Inc., 190 North Independence Mall West, Suite 601, Philadelphia, PA 19106, USA

Supplementary MATLAB programs to accompany this book are available at https://iopscience.iop.org/book/978-0-7503-3791-5.

Contents

Preface		ix
Acknowledgements		x
Author biography		xi

1	**Introduction**	**1-1**
1.1	A note of caution: rounding errors	1-2
1.2	More on the limitations of digital computers	1-4
	Exercises	1-5

2	**Introduction to programming with MATLAB**	**2-1**
2.1	Computer programming	2-1
2.2	Good programming practices	2-2
2.3	Introduction to MATLAB	2-2
2.4	HELP on MATLAB	2-5
2.5	Variables	2-5
2.6	Mathematical operations	2-7
2.7	Loops and control statements	2-9
2.8	Built-in MATLAB functions	2-9
2.9	Some more useful MATLAB commands and programming practices	2-11
2.10	Functions	2-12
2.11	Using MATLAB for visualisation	2-14
2.12	Producing sound using MATLAB	2-21
	Programming exercises	2-27

3	**Finding the roots and zeros of a function**	**3-1**
3.1	The roots of a polynomial	3-1
3.2	Graphical method	3-1
3.3	Solution of equations by fixed-point iteration	3-5
3.4	Bisection	3-7
3.5	Descartes' rule of signs	3-9
3.6	The Newton–Raphson method	3-10
3.7	The false position method	3-13
3.8	The secant method	3-14
3.9	Applications of root finding in physics	3-14
3.10	The finite potential well	3-15

3.11 The Kronig–Penney model   3-19
       Exercises   3-21

4   Interpolation   **4-1**

4.1 Lagrangian interpolation formula   4-1
4.2 The error caused by interpolation   4-4
4.3 Newton's form of interpolation polynomial   4-5
       Exercises   4-6

5   Numerical linear algebra   **5-1**

5.1 Solving a system of equations: Gaussian elimination   5-1
5.2 Evaluating the determinant of a matrix   5-6
5.3 LU decomposition   5-8
5.4 Determination of eigenvalues and eigenvectors: the power method   5-12
5.5 Convergence of the power method   5-14
5.6 Deflation: determination of the remaining eigenvalues   5-15
5.7 Curve fitting: the least-squares technique   5-17
5.8 Curve fitting: the generalised least-squares technique   5-19
       Exercises   5-20

6   Numerical integration and differentiation   **6-1**

6.1 Numerical differentiation   6-1
6.2 The Richardson extrapolation   6-3
6.3 Numerical integration: the area under the curve   6-4
6.4 Simpson's rules   6-7
6.5 Comparison of quadrature methods   6-11
6.6 Romberg integration   6-11
6.7 Gaussian quadrature   6-12
6.8 Gaussian quadrature for arbitrary limits   6-15
6.9 Improper integrals   6-19
       6.9.1 Limit comparison test   6-19
       6.9.2 Direct comparison test   6-20
6.10 Approximate evaluation of integrals using Taylor series expansion   6-23
6.11 The Fourier transform   6-24
6.12 Numerical integration using MATLAB   6-28
       Exercises   6-28

7	**Monte Carlo integration**	**7-1**
7.1	Error in multidimensional integration	7-1
7.2	Monte Carlo integration	7-2
7.3	Error estimate for Monte Carlo integration	7-6
7.4	Importance sampling Monte Carlo	7-8
7.5	The Box–Muller method	7-10
7.6	The Metropolis algorithm	7-11
7.7	Random number generators	7-12
7.8	The linear congruential method	7-14
7.9	Generalised feedback shift register	7-15
	Exercises	7-15

8	**Applications of Monte Carlo methods**	**8-1**
8.1	Random walks	8-1
8.2	The Ising model	8-4
8.3	Percolation theory	8-10
8.4	Simulated annealing	8-13
	Exercises	8-16

9	**Ordinary differential equations**	**9-1**
9.1	Differential equations in physics	9-1
9.2	The simple Euler method	9-4
9.3	The modified and improved Euler methods	9-8
9.4	Runge–Kutta methods	9-10
9.5	The Taylor series method	9-13
9.6	The shooting method	9-15
9.7	Applications to physical systems	9-17
	Exercises	9-22

10	**Partial differential equations**	**10-1**
10.1	Partial differential equations in physics	10-1
10.2	Finite difference method for solving ordinary differential equations	10-1
10.3	Finite difference method for solving PDEs	10-3
10.4	A finite difference method for PDEs involving both spatial and temporal derivatives	10-8
	Exercises	10-13

11   Nonlinear dynamics, chaos, and fractals   11-1

11.1 History of chaos   11-1
11.2 The logistic map   11-5
11.3 The Lyapunov exponent   11-13
11.4 Differential equations: fixed points   11-18
11.5 Fractals   11-21
      Exercises   11-28

Appendix A: Solutions to selected exercises   A-1

Preface

Computers have invaded every aspect of our lives. From the common man to the scientist, everyone now needs a minimal level of computer literacy. For most scientists and engineers, a basic familiarity with using and programming computers is almost essential. So much so, that if Descartes were alive today, he probably would have said, 'I compute, therefore I am'.

In past times, experimentation and theoretical investigation were the two ways of doing research. Computation has now become the third way of doing physics, and in many instances it complements experiment and theory. Computation can not only be used as a tool for research in physics, but it can serve as an invaluable tool for a student to learn physics. In order to write a program to simulate a physical system, a student needs to have a very solid understanding of the underlying physics. This solid understanding will help the student to break down the problem into small parts and instruct the computer to perform the simulation.

There are many books on computational physics or computational methods. Do we need one more? Books on computational physics describe numerical methods which even date back to the time of Isaac Newton. So there are a large number of books on numerical methods and numerical analysis. This book differs from earlier books in its emphasis on various topics and tries to strike a balance between the analysis of numerical methods and their implementation using programs.

The MATLAB software package is widely used by many scientists and engineers. The various toolboxes which are part of MATLAB, immensely simplify the task of writing programs to implement numerical methods. However, in this book, I have tried to emphasise the underlying algorithms and eschewed the use of built-in MATLAB functions, which render the implementation opaque to the user. Built-in MATLAB functions are also mentioned along with the discussion of the underlying algorithms.

This book is based on a one-semester course I have been offering to students of physics for many years. I have tried to select topics which would help students in the other courses they typically study, namely, classical mechanics, quantum mechanics, statistical mechanics, electromagnetic theory, and solid-state physics. In each chapter, the application of numerical methods to various areas of physics is mentioned immediately after the discussion of methods, so that the reader can easily grasp the relevance of the numerical methods and be motivated to study them.

I suspect that some instructors may find one semester insufficient to deal with all the topics included here. In such cases, the last chapter (nonlinear dynamics and chaos) or chapter eight (which deals with application of Monte Carlo methods) can be excluded. Exercises are given at the end of each chapter and students should diligently strive to solve them before verifying the solutions.

Programming assignments are also given along with the exercises. Some of these have hints for developing the program. The author can be reached through email at thiru@hyderabad.bits-pilani.ac.in. He welcomes suggestions and comments regarding the book. Additional material and programs related to this book can be accessed from https://universe.bits-pilani.ac.in/Hyderabad/pkthirivikraman/Publications.

Acknowledgements

Writing a book requires access to resources and a conducive atmosphere. I whole-heartedly thank Birla Institute of Technology and Science for providing me with the resources and time for writing this book. This book was possible as I was allowed to teach a course on computational physics for the last five years. Feedback from my colleagues and students enabled me to tweak the contents of this book to make it more accessible and reader-friendly. I thank my colleagues and students for their suggestions and comments. Encouragement provided by my family members was very crucial for the completion of this book. I thank IoP publishing and John Navas in particular, for giving me an opportunity to write another book, and the painstaking efforts of the editorial staff minimized the errors in this book.

Author biography

P K Thiruvikraman

P K Thiruvikraman is currently a professor of physics at the Birla Institute of Technology and Science, Pilani (Hyderabad Campus). He has more than two decades of experience in teaching courses in many areas of physics. He obtained his Master's degree in physics from the Indian Institute of Technology, Madras, and a PhD in Physics from Mangalore University. During his teaching, he has come to realise that most subjects that are typically offered by different departments are intimately interrelated. He tries to convey this to students by quoting examples from different areas of science and engineering. An avid reader of both fiction and non-fiction and a sports and movie enthusiast, he also tries to use these to enliven his classes. He has also authored a book titled *A Course on Digital Image Processing with MATLAB*.

Computational Methods Using MATLAB®
An introduction for physicists
P K Thiruvikraman

Chapter 1

Introduction

Traditionally, physicists were classified into those who do theoretical physics and those who do experimental physics. Experimental physicists perform experiments to study natural phenomena under carefully controlled conditions in the laboratory. Theoretical physicists come up with models and theories to account for the results of experiments and also work out the consequences of their theories. A third way of doing physics, namely, computational physics, has gained prominence in the last few decades. Interestingly, computational physics has features of both experimental and theoretical physics. In computational physics, we attempt to solve differential and algebraic equations using numerical techniques. In that sense, it is like theoretical physics. We can also simulate a system based on the underlying physical laws, which is like doing a virtual experiment. Hence, computational physics has something in common with theoretical as well as experimental physics.

Based on introductory courses in physics, you might have concluded that physics is mostly about solving differential equations for specific situations. Examples of differential equations we come across in various branches of physics are Newton's second law and the Euler–Lagrange equation in classical mechanics and the Laplace equation, the Poisson equation, and Maxwell's equations in electrodynamics. We can also add the Schrödinger equation, the wave equation, and the diffusion equation to that list.

What do you think is common to all these equations drawn from the various branches of physics? A little thought will convince you that all of them are linear, second-order differential equations (most of them are partial differential equations).

At this stage, you might ask whether nature is inherently linear? Definitely not. It is well known that we end up with nonlinear equations in most real-life situations, which is why the weather is so difficult to predict. Many nonlinear systems can behave chaotically (we will make an extensive study of chaos in chapter 11).

Classical (pre-20th century) physics avoided any discussion of nonlinear systems. Physicists and mathematicians spent a lot of time making approximations that led to

doi:10.1088/978-0-7503-3791-5ch1

linear equations. For example, consider the case of projectile motion as taught at the school level. All textbook problems ask us to ignore air resistance. Of course, in real life, the inclusion of air resistance is critical to determining the trajectory of a projectile. Army commanders would like to calculate the precise trajectory of a cannonball before firing it so that it can hit the intended target. If most real-life systems are nonlinear, why do we make approximations (such as ignoring air resistance) and linearize the equations describing the system? There are at least two fundamental reasons for this: (i) mathematical beauty: in the case of air resistance, the trajectory of a projectile would not be a beautiful parabola. Also, a nonlinear differential equation would not satisfy a very useful principle, such as the super-position principle, which is the cornerstone of optics. (ii) Analytical solutions: traditionally, mathematicians and physicists were interested in what are called 'closed-form' or 'analytical' solutions to differential equations. An 'analytical' solution means that the solution to a differential equation should be expressed in terms of 'known' functions, such as trigonometric functions or the logarithmic or exponential functions. However, this is not a precise definition, as one can always define a new function to be the solution to a particular differential equation. Many of the well-known special functions of mathematical physics, such as the Bessel function, are defined as the solutions to particular differential equations. Note that even if we obtain an 'analytical solution', we will need to numerically evaluate the value of the function at some particular point. Hence, the distinction between an 'analytical' and 'numerical' solutions is not as rigid as it is customarily made out to be.

The need to model systems realistically has meant that scientists and engineers have to deal with nonlinear differential equations that can be solved only by numerical methods, which leads to the importance of computational methods in present-day scientific research. To compete in the present-day world (in any field), one needs to compute.

There is another reason why computational methods are important. The physicist Richard Feynman once said: 'If you can't explain it to your grandmother, you don't understand it well enough.' I will paraphrase this and say that if you can't explain it to a computer, i.e. through a program, you do not understand it well enough. A computer will implement only what you have instructed it to do. Of course, you have to break up a calculation based on some theory into a sequence of operations (additions, multiplications, comparisons, etc.) so that the computer can follow it precisely. Any defect in your understanding will be exposed by the non-physical results produced by the computer.

A student can also test his/her understanding of physics by writing programs to model physical systems. To paraphrase an old Chinese saying: 'I hear a lecture and I forget; I see a demonstration and I remember; I compute and I understand'.

1.1 A note of caution: rounding errors

Over the last few centuries, the rapid development of mathematics and the physical sciences has meant that almost all problems that admit an analytical solution have been solved. Most problems that physicists and engineers deal with in present-day research require numerical methods for their solution. The advent of high-perform-ance computing has made it possible for most of these problems to be solved by numerical techniques. We have come a very long way since the time of P A M Dirac, who famously said (in 1929), *'The underlying physical laws necessary for the mathematical development of a large part of physics and much of chemistry are completely known, and the difficulty is only that the exact application of these laws leads to equations much too complicated to be soluble.'*

While it is true that numerical techniques can solve almost all problems, we should be cautious about the results produced by programs. While any program free of syntactic errors will always produce an output, this does not guarantee the correctness of the results.

We should, at all times, be aware of the limitations of computers while using numerical techniques.

We now present a few examples to illustrate the limitations of computers.

Example 1.1

Choose a number between zero and one. Multiply it by two and take only the decimal part of the number. Multiply the decimal part by two and repeat the operation. Store the results you get at each iteration. Do you always end up with zero? Why?

For instance, if we choose 0.6 as the initial number, we get 1.2 after multiplying by two and the decimal part is 0.2. On repeated multiplication, by 2, we get: 0.4, 0.8, 0.6, 0.2.... we see that the pattern should repeat after every four numbers. Try this on a computer, and you would be surprised by the results.

For the first forty iterations or so, the pattern repeats, but at some point, you start seeing numbers such as 0.6001, 0.2002, 0.4004 ... and after some time, the number simply becomes zero. What is happening here? Well, the results should not surprise you if you have understood the nature of digital computers. Computers use the binary number system, and therefore, all integers are represented as the sum of powers of two. In contrast, the fractional part of a number is represented as the sum of inverse powers of two. So 0.5 is represented in binary as 0.1000, while 0.25 is represented as 0.0100. A little thought will tell you that only some numbers can be expressed in this fashion using a finite number of bits. Numbers such as 0.5, 0.25, 0.125, 0.0625, 0.03125, 0.75, 0.625, 0.875... can be represented as sums of inverse powers of two, while numbers such as 0.1, 0.2, 0.4, and 0.6 cannot be represented exactly like this using the 64 or 32 bits that your machine has.

To realise why a number like 0.6 cannot be represented exactly in base two, imagine a computer that uses only four bits to represent such a number. What will be the representation of this number using four bits? Since the number is greater than 0.5, the first bit after the decimal point will be one. Now subtract 0.5 from 0.6. Since the difference is 0.1, which is less than 0.25, the next bit is zero. It is also less than 0.125, which means that the third bit is also zero. The fourth bit should be one, since 0.1 is greater than 0.0625. Therefore, using four bits, 0.6 is represented as 0.1001— but 0.1001 is actually $0.5 + 0.0625 = 0.5625$. The error in our representation is 0.0375. So when you type 0.6 (and you see 0.6 on the screen), remember that the computer is actually fooling you! It is showing 0.6 on the screen, but the actual number in the memory of the computer is 0.5625. In case you think the representation of 0.6 is inexact only because you used four bits, try using 8, 16, 32, or 64 bits[1]. Since the decimal-to-binary conversion becomes tedious with an increasing number of bits, we urge you to write a program to implement this conversion.

The above example should have given you a fair idea about the rounding errors present in the binary representation of numbers. With 64 bits, these errors are less than $1/2^{64}$, but nonetheless they exist.

A 64-bit computer cannot use all its bits to represent the digits of a decimal number, as explained below. A number such as 0.6 is known as a floating-point number in the language of computer science. In the floating-point[2] representation, 25.63 is represented as 0.2563×10^2. Note that this is slightly different from standard scientific notation, in which 25.63 would be represented as 2.563×10^1. Most computer systems and software packages (including MATLAB) use the Institute of Electrical and Electronics Engineers (IEEE) 754 standard to represent floating-point numbers. This standard uses 32 bits for a single-precision floating-point number and 64 bits for double precision. Out of the 64 bits available for the double-precision representation, one bit is used for the sign, 11 bits are used for the exponent, and the remaining 52 bits are used for the number. A preferred approach, called two's complement, incorporates the sign into the number's magnitude. Type the commands 'realmax' and 'realmin' at the command prompt to get the largest and smallest numbers, respectively, which MATLAB can represent[3].

With double precision, the rounding errors are small, but can be significant in some situations, as in example 1.1.

Now that we have discussed the binary representation at some length, you should be able to understand the results we obtained in example 1.1. Multiplying a number by two shifts all the bits to the left. The rightmost bit is zero after the multiplication. Therefore, repeated multiplication by two will create many zeros, and very soon, the entire number becomes zero (this is the case because the integer part is removed after each multiplication).

In addition to the rounding errors, we will see that each of the algorithms we use comes with its own error. The total error in a calculation will be the sum of these two errors. We will endeavour to reduce the total error to a minimum.

1.2 More on the limitations of digital computers

Example 1.1 and the subsequent discussion presented earlier illustrated the limitations that arise from the finite precision of computers. While we can come up with many examples from various fields, I will confine myself to one example drawn from the summation of series.

---

[1] The eight-bit representation of 0.6 is 0.10011001, which is actually equal to 0.59765625. Using 16 bits, we get to 0.5999755859375.

[2] It is known as 'floating-point' representation, as the point floats to the top (or to the left of the most significant digit).

[3] In the 2019 version of MATLAB, realmax gives 1.7977e+308, while realmin gives 2.2251e−308.

Mathematicians have come up with various techniques to determine the sum of many series. You have doubtless come across formulae for the sums of the geometric, arithmetic, and arithmetic–geometric series.

While analytical techniques exist to find the sums of a large variety of series, let us pretend for the time being that we are ignorant of those techniques and would like to calculate the sum of series numerically.

Suppose that we start with the following:

$$S_1 = \sum_{n=0}^{\infty} 2^{-n} = 1 + \frac{1}{2} + \frac{1}{4} + \dots.$$

You must know that this is a geometric series with a common ratio of ½ and that the sum of this series is two.

How would we evaluate the sum numerically? Since we cannot deal with an infinite number of terms, we can sum the series up to a finite number of terms. This would not matter much, as the terms become smaller and smaller for a convergent series such as the one given above.

The sum of the first two terms is 1.5, while the sum of the first ten terms is 1.999. The sum reaches the value of two within just fifty terms.

However, now consider a different series, as shown below:

$$S_2 = \sum_{n=1}^{\infty} \frac{1}{n} = 1 + \frac{1}{2} + \frac{1}{3} + \dots.$$

If we want to check the convergence of this series 'experimentally', we would sum the series up to a certain number of terms and check whether the sum converges to a finite value as we increase the number of terms.

The first ten terms of this series sum to 2.92, and the series increases very slowly. Even if we include the first billion terms, the sum is only 21.4116. However, beyond a certain number of terms, the series does not increase, as all terms beyond this point are too small and are rounded to zero, because of the finite precision.

In fact, the series S_2 is the famous harmonic series, which is known to be divergent[4]. This shows one limitation of digital computers. They can provide the sum of a convergent series, but in at least some cases, they cannot tell you whether a series is convergent or divergent.

Exercises:

1.1. Write a program to determine the binary representation of a number between zero and one.

1.2. Write a program to evaluate the sum of the following series:

---

[4] The book *Gamma: Exploring Euler's Euler's Constant* by Julian Havil, Princeton University Press, Princeton, NJ, provides a detailed historical background for the harmonic series together with proofs of its divergence.

$$S_3 = \sum_{n=1}^{\infty} \frac{1}{n^2} = 1 + \frac{1}{2^2} + \frac{1}{3^2} + \dots$$

This is a convergent series which can be summed analytically. Can you guess the analytical value from the results of your program?

1.3. Type '0.4-0.5+0.1' at the command prompt and note the result. Why do you get a non-zero value? Now type '0.1-0.5+0.4' and observe the result. Is it equal to zero?

Chapter 2

Introduction to programming with MATLAB

2.1 Computer programming

As we mentioned earlier, this is the age of computers, and a physicist is expected to have a reasonable degree of familiarity with computer programming. We will confine ourselves to programming with MATLAB. However, the syntax of MATLAB is quite similar to those of many other programming languages, and it should be possible for you to 'translate' a MATLAB program into C or Python. MATLAB has a toolbox named 'MATLAB coder', which can generate a C program from a MATLAB script.

I will assume that you are familiar with at least one programming language. If not, you should pick up the basics of programming by looking at the programs we discuss.

The crux of computer programming is to break up a given task into individual operations, each of which can be represented as a mathematical operation, such as multiplication, division, addition, subtraction, exponentiation, etc.

While mathematical operations can be performed by computers, we should also be aware of the subtle differences in notation between the language of mathematics and those used by computers.

If you type $A = 5$ at the MATLAB command prompt or as part of a program, the computer assigns the value 5 to the variable A. If, in a subsequent step, you type $A = A + 5$, the A on the right-hand side has the old value of A, i.e. five, but A is now assigned the value ten (the old value plus five). Note that the equation $A = A + 5$ is nonsense as far as mathematics is concerned, but it makes perfect sense to a computer. This is because $A = A + 5$ or any similar statement made in a computer program is not an equation in the usual sense of mathematics, but is to be interpreted as an assignment operation.

doi:10.1088/978-0-7503-3791-5ch2

2.2 Good programming practices

Many students are under the impression that pen and paper are redundant once we have access to a computer. However, you will soon discover that it makes sense to flesh out your algorithm on paper before you start typing a program. You will waste a lot of time if you create a program without a proper plan. Once you have a clear idea of the algorithm you are planning to follow, put it down on paper in the form of a flowchart or pseudocode.

Some good programming practices are listed below:

    (i) Give proper and meaningful names to the variables used to represent mass, velocity, etc. For example, m for mass and v for velocity.

    (ii) Do not hard code, i.e. if a parameter is used in many places in the program, its numerical value should be entered only once. Use the name of the parameter in the rest of the program.

    (iii) Think about the problem before you start to write the program.

    (iv) Use pen and paper before you touch the keyboard!

    (v) Design down: use functions (more about functions later on in this chapter).

    (vi) Write comments within the program so that the logic of the program is straightforward.

    (vii) If you are using MATLAB, use the 'clear' command at the beginning of a session to clear old data from the memory. This is especially required when you are using a shared computer or want to remove old data.

With this one-page introduction to programming, let us now get on with using MATLAB. On the way, we will also cover the basics of programming.

2.3 Introduction to MATLAB

MATLAB[1] is one of the software packages most widely used to analyse and visualise data in science and engineering. MATLAB contains many built-in functions that implement many of the algorithms we will discuss in this book. Mathematica, which was initially developed for symbolic computation, also offers many in-built commands for numerical computation.

While MATLAB is a very user-friendly package, it may take some time for a new user to get used to it. This short introduction is intended to familiarise the reader with the MATLAB environment and the commands required to run the programs listed in this book.

We will assume that you have installed MATLAB on your computer. Once MATLAB is installed, click on the MATLAB icon on the desktop to run MATLAB.

Once the icon is clicked, the entire MATLAB package with all the toolboxes (installed on your computer) is loaded into the computer's memory. Most of the screenshots (figures 2.1–2.3) pertain to version R2019b of MATLAB. If you are

---

[1] MATLAB is a product of Mathworks. Complete details about MATLAB can be obtained from http://www.mathworks.in/products/matlab/.

Figure 2.1. MATLAB icon; R2019b is the version of MATLAB used while writing this book.

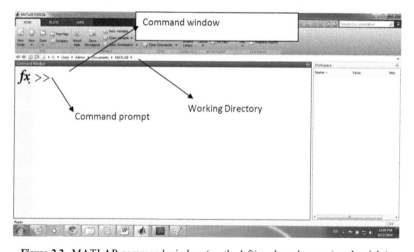

Figure 2.2. MATLAB command window (on the left) and workspace (on the right).

using a different version of MATLAB, the appearance of the windows may be slightly different. However, the commands and programs given here should work as intended, even if you are using another version of MATLAB.

MATLAB commands can be typed directly on the 'command window' which opens when MATLAB is loaded.

Figure 2.3. Close-up of the MATLAB command window. The current working directory is displayed above the command window.

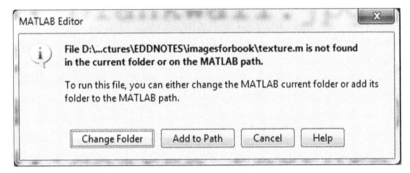

Figure 2.4. Message from MATLAB which is displayed if you attempt to run a program which is not in the current folder (directory).

The command window is shown in figure 2.2. The 'workspace', which is a separate window shown in figure 2.2 (it is to the right of the command window in figure 2.2) displays the variables currently available.

Programs can be executed either by typing the name of the program file (without the '.m' extension) at the command prompt or by clicking the 'Run' icon when a '.m' file is open.

New '.m' files can be created by clicking on 'New Script' in the 'HOME' tab (figure 2.3).

While running MATLAB programs ('.m' files), ensure that the image or text files required by the program are stored in the directory in which the program is stored. The current working directory is displayed above the command window (see figures 2.2 and 2.3). If you wish to access a file in another directory, the entire path has to be specified. Before running a program, you have to change the working directory to the directory in which the program is stored. In recent versions of MATLAB, you will be prompted to change the directory when you attempt to run a program that is outside the working directory. You can choose the 'Change Folder' option if such a message appears (figure 2.4).

If you attempt to access a file which is not in the current directory, MATLAB will display an error message (figure 2.5)

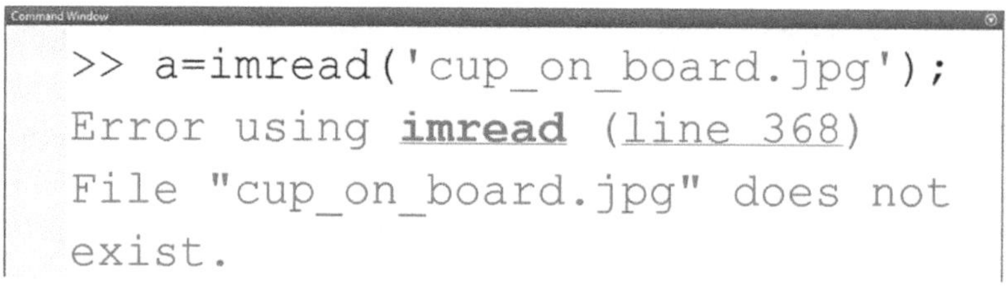

```
>> a=imread('cup_on_board.jpg');
Error using imread (line 368)
File "cup_on_board.jpg" does not
exist.
```

Figure 2.5. Error message displayed if an image file is not in the current folder.

2.4 HELP on MATLAB

There are some excellent books on using MATLAB. The following is a short list of the many books available for this purpose:

1. *Getting Started with MATLAB*, Rudra Pratap, Oxford University Press;
2. *Mastering MATLAB*, Duane C Hanselman and Bruce L Littlefield, Pearson;
3. *MATLAB: A Practical Approach* by Stormy Attaway, Butterworth-Heinemann (Elsevier), 2009;
4. *Numerical Computing with MATLAB*, Clive Moler, available online at: http://www.mathworks.com/moler/chapters.html.

You can also get help on the syntax of any particular command by typing 'help' followed by a space and the name of the command in the command prompt. For example, typing 'help plot' will display the syntax for the plot command (which is used to plot two-dimensional graphs).

Typing 'help' by itself displays all the help topics. For example, some of the help topics are 'general', 'elmat', and 'elfun'. Typing 'help general' will display all the general-purpose commands that are part of MATLAB, while typing 'help elmat' will display elementary matrices and matrix manipulation. Typing 'help elfun' displays all the elementary math functions that are part of MATLAB.

You can then choose any of the commands or functions and type help followed by that command, as described earlier.

2.5 Variables

The syntax of MATLAB is very similar to that of the C programming language, but one major advantage of MATLAB is that, unlike C, one need not define variables before using them.

The basic data type of MATLAB is an array. Even scalars are treated as arrays of size 1×1. There is no need to specify the dimensions of the array at beginning of the program. MATLAB automatically identifies the type of array being used. However, for efficiency and speed of implementation, it is suggested that we initialise and hence specify the dimensions of the array.

Unlike C, array indices start with one.

To input a matrix, say

$$A = \begin{bmatrix} 1 & 2 & 5 \\ 3 & 9 & 0 \end{bmatrix}$$

We type: $A = [1\ 2\ 5;\ 3\ 9\ 0]$ on the command prompt (or in a '.m' file).

The semicolon indicates the end of a row; hence, $u = [1\ 3\ 9]$ produces a row vector and $v = [1;\ 3;\ 9]$ produces a column vector.

Matrix elements can be referred to as $A(i,j)$. $A(i,j)$ refers to the element in the ith row and jth column of matrix (array) A.

$A(m:n,k:l)$ — refers to rows m to n and columns k to l.

$A(:,5:20)$ — refers to the elements in columns 5 through 20 of all the rows of matrix A.

While assigning names to variables (and also '.m' files), it should be remembered that the names of variables and files (or functions) have to begin with a letter of the alphabet. After that, they can contain letters, digits, and the underscore character (e.g., value_1), but cannot contain a space. MATLAB is case-sensitive. This means that there is a difference between upper- and lowercase letters. Therefore, the variables mynum, MYNUM, and Mynum are all different. We can use the command 'who' to obtain the names of variables currently loaded into the MATLAB's memory.

Sometimes, we may wish to delete certain variables from MATLAB. In such a case we can use 'clear variable name'. Typing

>> clear Mynum

will clear (delete) Mynum from memory, while typing

>> clear

will delete all variables from memory. It is a good practice to use 'clear' at the beginning of each program, so that we do not unwittingly re-use the values of some old variables. Of course, if we want to access a variable stored in memory, then we should not use this command.

When assigning names to variables, it is better to avoid names that are used for 'built-in' functions. For example, 'i' is reserved for the square root of -1 in MATLAB. If we use it to refer to another variable (for example, if we type '$i = 2$' at the command prompt or in a MATLAB program), then the MATLAB name 'i' will no longer refer to $\sqrt{-1}$, but will be treated as being equal to the value which we have assigned. MATLAB also understands 'pi' as being equal to 'π'. It is a good practice to use 'pi' in MATLAB programs instead of '22/7' or '3.14' in our programs. Using 'pi' will improve the accuracy of our computations. Of course, we should remember that we should not use 'pi' to refer to any other variable. You should especially keep the above-mentioned points about 'i' and 'pi' in mind when writing programs to compute a Fourier transform (for instance), as Fourier transforms involve both 'π' and $\sqrt{-1}$. Note that we can also use 'j' instead of 'i' to refer to $\sqrt{-1}$ in MATLAB.

The size of a matrix can be obtained using the command

$s = \text{size}(A)$

$s(1)$ contains the number of rows and $s(2)$ contains the number of columns.

The size command can be used to obtain the numbers of rows and columns in an array. If $s(1)$ represents the number of rows and $s(2)$ the number of columns (as given above), then we can use $s(1)$ and $s(2)$ in the rest of the program to refer to the numbers of rows and columns of that array instead of hard-coding these numbers.

2.6 Mathematical operations

Most mathematical operations are represented in the usual fashion in MATLAB, for example:

Addition	+
Subtraction	−
Multiplication	*
Division	/
Exponentiation	^

If A and B are matrices, then A*B will multiply the matrices A and B. Sometimes, we need to multiply the corresponding elements in two matrices; we would then represent the multiplication operation as 'A.*B'.

The '.' signifies that the first element in A has to be multiplied by the first element in B and so on.

To summarise, if we have to perform matrix multiplication, we use the expression A*B. If we use the expression A*B, then the number of columns in A has to be equal to the number of rows in B. If we want to multiply the corresponding terms in A and B, the two matrices should have equal numbers of rows and columns.

A '.' is used whenever we need to perform term-by-term operations:

Multiplication	.*
Division	./
Exponentiation	.^

When performing mathematical operations, such as multiplication, division, addition, subtraction, etc. MATLAB uses the well-known 'BODMAS' rule.

This rule is a convention which tells us the order in which these operations have to be performed.

'BODMAS' is actually an abbreviation for:

Brackets first (**B**), which implies that we compute the result of operations within a pair of brackets first

Order (**O**)—orders, i.e. powers and square roots are computed next

Division and multiplication (**DM**) are performed next; both are on an equal footing, so the expression has to be evaluated from left to right

Addition and subtraction (**AS**)—again, both addition and multiplication are on an equal footing, and the expression is to be evaluated from left to right

Relational operations are represented in a similar manner to that of C with minor variations:

<	Less than
<=	Less than or equal to
>	Greater than
>=	Greater than or equal to
==	Equal
~=	Not equal

A single '=' (equal to) is an assignment operation, whereas '==' is used to check whether the variables on either side of this symbol are equal to each other.

Apart from the relational operators, MATLAB also has logical operators. Some of these operators are mentioned below:

&&	Logical AND
||	Logical OR
~	Logical NOT
XOR	Exclusive OR

MATLAB supports a number of data types:
- int (integer)—within this class, we can have int8, int16, int32, int64, and uint8. The numbers following 'int' refer to the number of bits used to store the variable. We obviously need more bits to store larger numbers; 'uint8' means that the variable is stored as an unsigned 8-bit integer. This is the data type used to store the intensity values of pixels in an image. Unsigned integers are used, as the intensity values are all positive.
- float (real numbers)—single or double precision.
- characters or strings—the values of variables which are characters/strings are enclosed in single quotes. E.g. >> c = 'cash';

Sometimes, we may wish to take an input from the user at the beginning of a program. The 'input' command accomplishes this:

g=input('Enter a number : ')

On typing this command, the user is prompted to enter a number. The number that the user types will be stored in the variable g. The 'disp' command can be used to display the value of a variable. The value of a variable may also be displayed on the screen by simply typing the variable name without a semicolon at the end.

For example:

C = A*B

will calculate and display the value of the matrix C, while

C = A*B;

will calculate C, but will not display it on the screen.

2.7 Loops and control statements

Sometimes, we require the same operation to be repeated many times. For example, we need to evaluate the force on a particle at each time step. In such cases, we introduce 'loops' into our program.

The syntax used to create 'for' and 'while' loops is very similar to that of C. For example, if we need to add two to all the odd-numbered elements in a one-dimensional array A with 101 elements, we use the following commands:

```
for i = 1:2:101
A(i) = A(i)+2;
end
```

If it is necessary to add two to every member of the array A, the same may be accomplished without a 'for' loop. Simply type

```
A=A+2;
```

The general syntax of the 'for' loop is:

for intial value:increment:final value

If the increment is not specified in the 'for' loop statement, a default value of one is assumed by MATLAB.

All the commands that lie between a 'for' statement and its corresponding 'end' will be executed multiple times, as specified in the 'for' loop command. The 'for' and 'while' loops can be broken (when a certain condition is satisfied) using the 'break' command.

On some occasions, we may need to change the flow of a program based on whether or not a particular condition is satisfied. On such occasions, we use the 'if' condition to check whether the given condition is satisfied and accordingly change the flow of the program. The general form of the 'if' statement is:

```
IF expression
statements
ELSEIF expression
statements
ELSE
statements
END
```

We can use relational operators ('= =', '~ =', etc.) in conjunction with logical operators in the 'if' condition's expression. We can also use nested 'if' statements (an 'if' condition within another 'if' condition) in certain cases.

2.8 Built-in MATLAB functions

It is not within the scope of this book to list and discuss all the built-in functions available in MATLAB and its toolboxes. However, we discuss below a few of the built-in functions you are likely to use regularly:

- abs(x): calculates the absolute magnitude of x. For example if $x = -5$, abs $(x) = 5$;

    if x is complex then abs(x) will be equal to the modulus (magnitude) of the complex number.

- rem(x,y): computes the remainder after dividing x by y; mod(x,y) also gives the remainder after division. However, 'mod' and 'rem' give different results if x and y have opposite signs.
- Trigonometric functions: sin(x) calculates the sine of x, where x is assumed to be in radians. The cosine, tangent, cotangent, secant, and cosecant of an angle can be calculated in a similar manner.
- $a$sin(x) gives the inverse sine or arcsine of x in radians. All other inverse trigonometric functions can be calculated in a similar manner.
- sinh(x) gives the hyperbolic sine of x. All other hyperbolic functions may be computed in a similar manner.
- *rand or rand*() generates a pseudorandom number in the range 0 to 1; rand(M,N) generates an $M \times N$ matrix of random numbers in the range 0 to 1. rand will only generate a pseudorandom number and not a truly random number as a computer is a completely deterministic machine and it has been pointed out that it is impossible to obtain a string of truly random numbers using an instrument which is completely deterministic[2]. The generation of random numbers may be useful in various contexts: for example, when we want to simulate noise. If the distribution of the noise is non-uniform, we have to convert the uniform (random) distribution to the required distribution using the methods described in chapter 7.
- *min*(A): returns the minimum of a set of values in the vector A. If A is a matrix, min(A) returns a column vector. If we desire to obtain the minimum of a set of values in a matrix (for example, an image), we should modify this command to read min(min(A)).
- *max*(A): in a similar manner, max(A) returns the maximum of a set of values in the vector A. If A is a matrix, we should use max(max(A)) to obtain the maximum value of all the elements of the matrix A.
- *sum*(A): this built-in function returns the sum of all components of the vector A. If A is a matrix, sum(A) will return a vector whose components are the sums of the individual columns of A. If we require the sum of all the elements of matrix A, we need to modify this command to read *sum* (*sum*(A)).
- *Mean*(A) *and std*(A) return the arithmetic mean and the standard deviation of A, respectively. If A is a matrix, we need to modify the commands in a similar way to the cases of min, max, and sum.

---

[2] It is relevant at this point to quote the famous mathematician Jon von Neumann: 'Anyone who considers arithmetical methods of producing random digits is, of course, in a state of sin'.

2.9 Some more useful MATLAB commands and programming practices

We may sometimes need to read the values of variables from a file or write the values into a file. In such a case, we can use the commands 'fprintf' (to print into a file) and 'fscanf' (to read from a file). Before reading or writing from a file, we naturally need to open the file using the 'fopen' command. 'fopen' will create a file with the specified name in the current folder, if such a file does not already exist. An example is given below:

    fid = fopen('mydata.txt');
    A = fscanf(fid,'%d');
    fclose(fid);

Here, '%d' is used because the file 'mydata.txt' contains integers; 'fclose' is used to close the file after reading from it. Writing into a file can also be achieved as shown below:

    fid = fopen('test.txt','w');
    fprintf(fid, '%6.2f \n',y);
    fclose(fid);

In the above example, the command 'fprintf' stores the values of the variable 'y' in the file 'test.txt'; '%6.2f' refers to the fact that the data type of y is 'float', with a precision of two decimal places. Note that the MATLAB syntax for both 'fprintf' and 'fscanf' is very similar to the syntax for these commands in C.

MATLAB can also read data from a Microsoft EXCEL file. The command 'xlsread' reads data from an Excel file, while the command 'xlswrite' writes data from memory into an Excel file.

When writing programs, it is useful to write comments at appropriate locations in the program. These comments will help us understand the program's working and will be especially useful if we want to refer to the program after a long time has elapsed. In such cases, we have generally forgotten the logic behind the program!

Comments can be enclosed within single quotes, as shown below:

    'This is a comment';

If you want to comment out an entire section of the program, you can select that section and choose the 'comment' option after right-clicking the mouse (figure 2.6). Commenting out part of a program is a useful tool for debugging it. If a program has a bug, i.e. a logical flaw, due to which it does not give the expected output, then it is usually possible to isolate the flaw by 'commenting' out some part which we suspect of having the flaw. When we 'comment' out some part of the program, MATLAB will not execute that part. When we click the 'Run' button, MATLAB will only execute the remaining (uncommented) part of the program. If the remaining part of the program is executed flawlessly, then we have succeeded in isolating the 'bug'. As seen in figure 2.6, we can 'uncomment' this part of the program once we have fixed the bug.

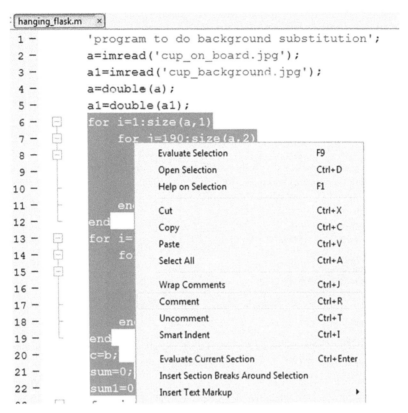

Figure 2.6. It is possible to 'comment' out entire parts of a program by selecting that part and right-clicking the mouse. A screenshot of this is shown here.

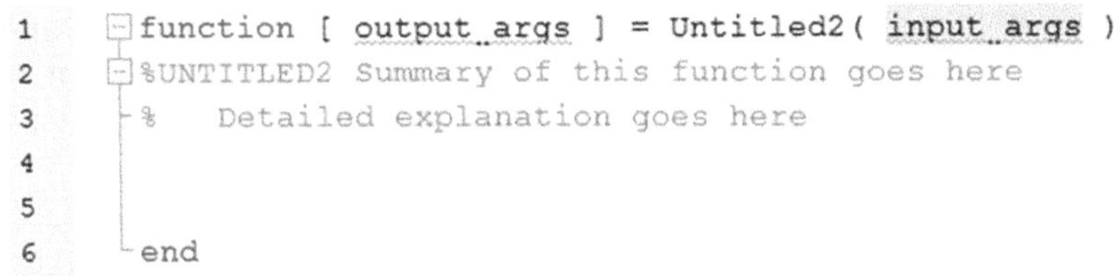

```
1    function [ output_args ] = Untitled2( input_args )
2    %UNTITLED2 Summary of this function goes here
3    %   Detailed explanation goes here
4
5
6    end
```

Figure 2.7. Definition of a function. You will see this if you click 'New function' in the HOME tab of MATLAB.

2.10 Functions

Like many programming languages, MATLAB allows the user to define his/her/ their own functions which can then be called from within other programs or functions.

Functions are also stored as '.m' files, but the first line in such a file should have the word 'function' followed by the name of the function. The input and output arguments should also be mentioned (figure 2.7). The input arguments are the

```
minmax.m    ×  readnumb.m    ×
1 -        fid=fopen('numb.txt','r');
2 -        s=fscanf(fid,'%d');
3 -        [mn,mx,m]=minmax(s);
4
```

Figure 2.8. The program 'readnumb.m', which reads the array of numbers in the text file and stores it in an array (in MATLAB). The program also calls a user-defined function.

```
minmax.m    ×  readnumb.m    ×
1     function [ mn,mx,m ] = minmax( a )
2     % Function to calculate the minimum, maximum and mean of a set of numbers;
3 -   mn=min(a)
4 -   mx=max(a)
5 -   m=sum(a)/size(a,1)
6
7
8
9 -   end
```

Figure 2.9. Example of a user-defined function in MATLAB. Note that it is best to make the name of the file the same as the name of the function.

variables that the function takes as an input to calculate the output variables (which are specified in the output arguments).

We give below a simple example of a function which can be defined and used in MATLAB. Let us assume that we have a one-dimensional array of numbers stored in a file named 'numb.txt'. Let us say that we are required to compute the minimum, maximum, and the mean of this array of numbers. We accomplish this by writing a program which reads the array of numbers from the text file and then calls a function which calculates the required quantities.

The program and the function are reproduced above (figures 2.8 and 2.9):

It can be seen from figure 2.9 that the user-defined function '*minmax*' calculates the minimum and maximum of the set of values by using the built-in functions *min*, *max*, and *sum*.

The advantage of defining a function is that if it is later necessary to compute the same quantities (the minimum, maximum, and mean) for another array named 'S' (for example), then we need not modify the definition of the function, but we only need to modify the argument of the function when it is called. When the function is called from the program 'readnumb.m', we replace 'minmax(a)' with 'minmax(S)' to calculate the minimum, maximum, and mean of S.

It is interesting to note that even the descriptions of user-defined functions can be accessed using the help command. In the example under consideration, typing 'help minmax', will cause the following lines to appear on the screen:

>> help minmax

Function to calculate the minimum, maximum, and mean of a set of numbers;

The short description of the function, which appears as a comment within the function, is be displayed when we type 'help <function name>' at the command prompt.

It is good programming practice to break up a lengthy program into many functions. This enables us to check each function separately for bugs and allows us (and others who happen to read the program) to understand the program's working.

When using functions, it should be kept in mind that only the variables which appear as the output arguments of the function can be accessed from outside the function. You can easily confirm this by deleting the left-hand side of the assignment operation in line three of readnumb.m (figure 2.8). If you do that, you will notice that you can no longer access those variables from outside the function. If you want certain variables to be accessible from outside the function, you should include them in the output arguments of the function or define them to be global variables.

It is standard programming practice to use recursion when writing programs. Recursion refers to calling a function from within itself. You can see that recursion is useful for the implementation of many algorithms, such as the fast Fourier transform (FFT). However, older versions of MATLAB have an upper limit on the number of times a function can be called recursively. Hence, it is better to avoid recursion when using MATLAB. It should be noted that a program which uses recursion can always be rewritten such that the recursion is replaced by iteration.

2.11 Using MATLAB for visualisation

As mentioned earlier, MATLAB is a powerful tool for scientific visualisation. A few ways in which MATLAB is useful for visualisation are listed below:

- Plotting graphs (2D and 3D);
- Vector fields, scalar fields;
- Surfaces, etc;
- Dynamic plots (Lissajous figures, phase-space portraits, trajectories of projectiles, etc).

Suppose that we wanted to plot y versus x, where $y = \sin(x)$. We would first need to decide the range of x over which we want to plot the function. If we want to plot the function for values of x ranging from -10 to $+10$, then the following lines of code will do the job.

$x = -10:0.1:10$;
$y = \sin(x)$;
$plot(x,y)$

The line $x = -10:0.1:10$; generates values of x (as a row vector) from -10 to $+10$ in steps of 0.1.

These lines of code generate figure 2.10.

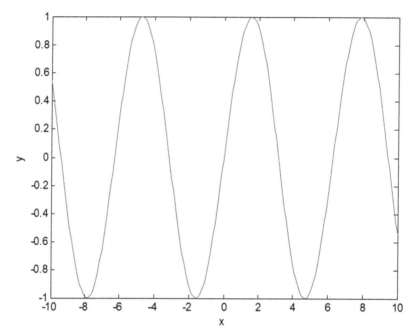

Figure 2.10. A plot of $y = \sin(x)$.

MATLAB can also generate surface plots for functions of two variables. For instance, let us say we wish to plot the function $y = \dfrac{1}{\sqrt{2}}(\cos(\pi x_1) \cos(2\pi x_2) + \sin(2\pi x_1) \sin(\pi x_2))$ which, incidentally, is the wavefunction for two particles confined in a box. The following lines of code will generate the plot for y as a function of x_1 and x_2.

```
u=-0.5:0.01:0.5;
[x1,x2]=meshgrid(u,u)
y=(1/2^0.5)*(cos(pi*x1).*cos(2*pi*x2)+sin(2*pi*x1).*sin(pi*x2));
surf(x1,x2,y);
```

Figure 2.11 shows the surface plot produced by this code.

We sometimes want to visualise a particular charge density configuration or a temperature profile. This can also be done using MATLAB. Figure 2.11 shows a plot of the charge distribution $\rho(r, \theta) = k\dfrac{R}{r}(R - 2r)\sin\theta$.

Here, positive charge is shown in red and negative charge is shown in blue. As can be seen from the functional form of the charge distribution, the charge density is positive for $r = 0$ to $r = R/2$, while it is negative for $r > R/2$ to $r = R$. What is special about this particular charge distribution? It is a distribution for which the monopole (i.e. the total charge) and the dipole moments are both zero[3]. Looking at the charge distribution (figure 2.12) may help you to understand how or why both these

---

[3] See problem 3.26 of *Introduction to Electrodynamics* by D J Griffiths, 3rd edn, Pearson, London, 2003.

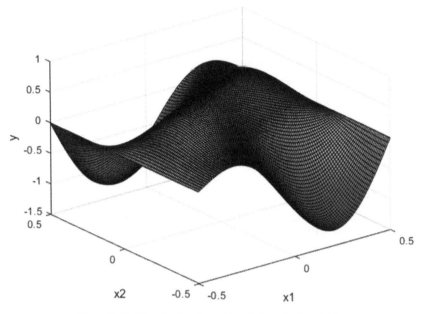

Figure 2.11. Plot of a function of two independent variables.

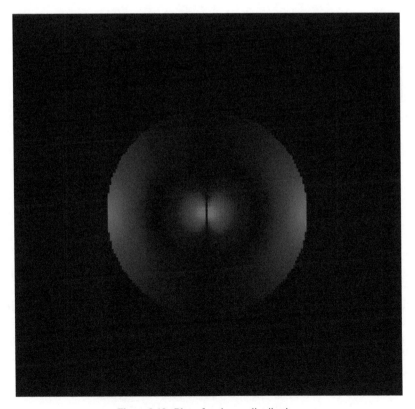

Figure 2.12. Plot of a charge distribution.

moments are zero. Interested readers can examine the MATLAB program given below for further details of the generation of such figures.

```
'plot of charge distribution';
clear;
clf
clc
R=50;
'initializing an array a to store the colours';
for x=1:200,
    for y=1:200,
        for k=1:3,
        a(x,y,k)=0;
    end
    end
    end
     'scanning the image of size 200 x 200 pixels and changing the
    RGB value of the pixel to red or blue according to the given
    charge density';
    m=1;
    for x=1:200,
        for y=1:200,
            theta(m)=atan2(y-100,100-x);
            theta(m)=abs(theta(m));
            m=m+1;
            r(x,y)=((x-100).^2+(y-100).^2).^0.5;
            if r(x,y)<50,
                t=(R/r(x,y)^2)*(R-2*r(x,y))*sin(theta(m-1));
                if t<0,
%                   t=t*50;
                    t=(255/log(256))*log(1+t);
                    a(x,y,3)=abs(t);

                else
                    t=(255/log(256))*log(1+t);
                    'log transformation to improve visibility';
                    a(x,y,1)=abs(t);
                end
            end
        end
    end

    a=uint8(a);
    imshow(a)
```

In various branches of physics and engineering, such as electrodynamics and fluid mechanics, we come across vector fields. Visualising these vector fields, which may be an electric field or a velocity field, will help us to understand concepts such as the curl and divergence of these fields in a better way.

Consider the vector field $\overline{V} = y\hat{i} + x\hat{j}$. This vector field is visualised in figure 2.13.

The MATLAB function 'quiver' was used to generate this plot. The program given below was used to generate this vector field:

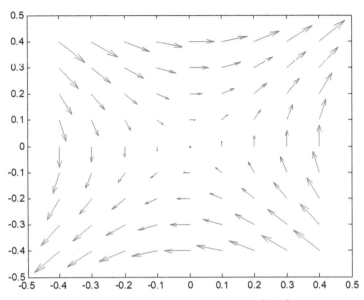

Figure 2.13. Plot of the vector field $\vec{V} = y\hat{i} + x\hat{j}$.

```
clf;
clear;
k=0;
l1=input('enter exponent of x for x component of vector
field');
m1=input('enter exponent of y for x component of vector
field');
l2=input('enter exponent of x for y component of vector
field');
m2=input('enter exponent of y for y component of vector
field');
c1=input('enter coefficient of x component of vector
field');
c2=input('enter coefficient of y component of vector
field');
 for i=1:9,
for j=1:9,
 k=k+1;
 x(k)=(i-5)/9;
 y(k)=(j-5)/9;
 u(k)=c1*x(k)^l1*y(k)^m1;
 v(k)=c2*x(k)^l2*y(k)^m2;;
end
end
quiver(x,y,u,v);
```

This program tries to plot a field of the form:

$$\vec{E} = c_1 x^{l_1} y^{m_1} \hat{i} + c_2 x^{l_2} y^{m_2} \hat{j}$$

Various functions can be tried by choosing suitable values for $c_1, c_2, l_1, l_2, m_1, m_2$, etc.

For example, setting c_1 equal to zero chooses a field which has only a y component. This expression should also clarify the important concept that the x component of a field can be a function of the y co-ordinate and vice versa.

For many problems, a dynamic plot, i.e. an animation, will also be helpful. This can easily be achieved in MATLAB by using the plot command within a FOR or WHILE loop. Each time the loop is executed, the plot is redrawn with the current values of the variables. This will be useful, for instance, in visualising the motion of a projectile or for generating Lissajous figures. The plot command should be followed by a 'pause' command. Without the pause command, the plot will be redrawn at a rapid rate, and we will not be able to follow the changes on the screen. The pause command slows down the execution of the program, allowing us to follow the motion on the screen. The time delay in seconds introduced by the pause command can be specified. For instance, pause (0.01) introduces a delay of 0.01 seconds, which is reasonable for most animations.

A program that generates a dynamic plot of a Lissajous figure is given below:

```
'superposition of SHMs';
clf;
axis([-1 1 -1 1]);
m=input('enter the total time(in seconds) for which you want to plot the
motion of the oscillators')
dt=input('enter time interval (dt) in seconds between one point which is
to plotted to be plotted and the next');
m=m/dt;
hold on;
x=0;
y=0;
t=0;

w1=input('enter angular frequency of SHM along x axis');
w2=input('enter angular frequency of SHM along y axis');
phi=input('enter phase difference in radians (enter pi instead of 3.14)
between the motion along x and y');
plot(x,y,'*');

  for i=1:m,
      t=t+dt;
      x=cos(w1*t);
      y=cos(w2*t+phi);
      plot(x,y,'*')
      pause(0.1);
  end
```

The following is a brief discussion of the theory of Lissajous figures: let us say that a particle is subjected to a simple harmonic motion with frequency ω_1 along the x-axis and frequency ω_2 along the y-axis. For simplicity, we consider the case in

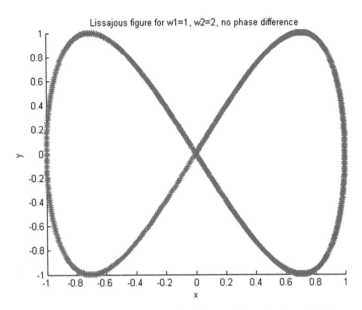

Lissajous figure for w1=1, w2=2, no phase difference

Figure 2.14. A Lissajous figure whose frequencies are in the ratio 1:2.

which the amplitudes of the two motions are equal (set equal to unity in the program).

The equations that represent the motions are as follows:

$x = \sin(\omega_1 t)$

$y = \sin(\omega_2 t + \varphi)$.

The equation which represents the trajectory of the particle can be obtained by eliminating t from the two equations. However, many interesting features of Lissajous figures can be appreciated by directly plotting the x and y coordinates as a function of t. Try varying the frequencies of simple harmonic motion along the x and y axes and also the phase difference between them and examine the figures generated by the program. You will develop an intuitive understanding of Lissajous figures by 'playing' with this simulation. When this program is run, the Lissajous figure is plotted continuously on the screen to make it more realistic. A few Lissajous figures are shown below (figures 2.14–2.16).

A particularly interesting exercise is to observe the trajectory for the case in which the ratio of the frequencies is an irrational number. In such a case, the curve will never close on itself. However, on a computer, since irrational numbers are represented by a finite number of decimal places, the trajectory will actually close after a large number of cycles. The picture given below is plotted for the case in which the frequency along the x-axis has been set to 22/7 and the frequency along the y-axis is one. Of course, this is an approximation of π. The student can use a better approximation for π and plot the resulting curve. This serves as a simple introduction to chaotic trajectories for the student. Note that in this case the curve does close on itself, since the ratio of the frequencies is a rational number (figure 2.17).

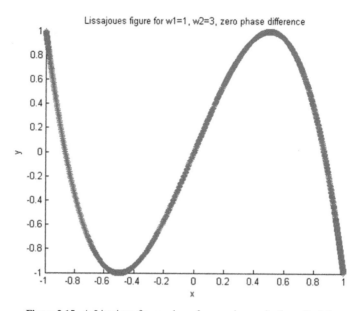

Figure 2.15. A Lissajous figure whose frequencies are in the ratio 1:3.

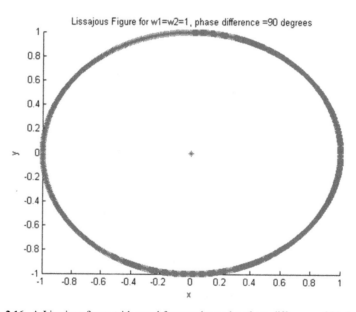

Figure 2.16. A Lissajous figure with equal frequencies and a phase difference of 90 degrees.

2.12 Producing sound using MATLAB

We conclude this chapter with an example that shows that sometimes hearing may be better than seeing. You might recall from elementary courses that superposing two sound waves of slightly differing frequencies leads to the production of beats (see figure 2.18).

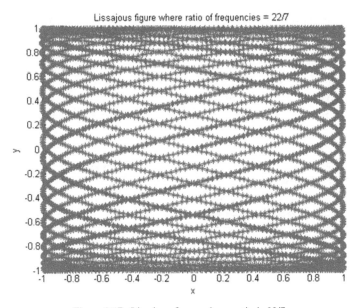

Figure 2.17. Lissajous figure whose ratio is 22/7.

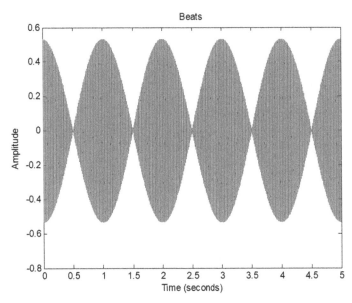

Figure 2.18. Beats produced when two sound waves that have frequencies of 300 Hz and 301 Hz are superposed.

In this case, I believe that hearing the sound gives us a far better idea of what is meant by 'beats', i.e., the variation in the intensity of the sound. MATLAB has some simple functions for producing sound. These functions can easily be modified to produce beats.

The program to generate sound of a particular frequency is given below:

```
'program to generate sound of different frequencies - sound files';
clear

duration = 10;          % duration in seconds
 amplitude = 0.8;          % amplitude
 f1 = input('enter the required frequency in Hertz');          % frequency in Hertz


phi = 2*pi*0.5;          % phase offset, e.g.: 2*pi*0.25 = 1/4 cycle
%% configure output settings
fs = 44100;              % sampling rate
T = 1/fs;                % sampling period
t = 0:T:duration;        % time vector
%% create the signal
omega1 = 2*pi*f1;        % angular frequency in radians

 signal = cos(omega1*t + phi)*amplitude;     % sinusoidal signal

%% plot the signal
plot(t, signal);
xlabel('Time (seconds)');
ylabel('Amplitude');
title('Beats');
%% play the signal
sound(signal, fs);

wavwrite(signal, fs, 'sound_freq.wav')
```

The program to generate beats is given below:

```
'program to generate beats - sound files';
clear

duration = 5;          % duration in seconds
 amplitude = 0.8;          % amplitude
 f1 = input("enter the frequency of the first SHM in Hz");          % frequency in Hertz
 f2 = input("enter the frequency of the second SHM in Hz");

phi = 2*pi*0.5;          % phase offset, e.g.: 2*pi*0.25 = 1/4 cycle
%% configure output settings
fs = 44100;              % sampling rate
T = 1/fs;                % sampling period
t = 0:T:duration;        % time vector
%% create the signal
```

```
omega1 = 2*pi*f1;          % angular frequency in radians
omega2 = 2*pi*f2;

partial1 = cos(omega1*t + phi)*amplitude;     % sinusoidal partial 1
partial2 = cos(omega2*t + phi)*amplitude;     % sinusoidal partial 2

signal = (partial1 + partial2)/3;
%% plot the signal
plot(t, signal);
xlabel('Time (seconds)');
ylabel('Amplitude');
title('Beats');
%% play the signal
sound(signal, fs);

wavwrite(signal, fs, 'beats_300_310.wav');
```

Listen to the beats (300 Hz and 301 Hz):

Audio 2.1. Frequencies 300 Hz and 301 Hz. Available at https://iopscience.iop.org/book/978-0-7503-3791-5.

The MATLAB program also gives you the freedom to enter different frequencies for the input. You will realise that the beats are less distinctly heard if the difference in frequencies is large.

For instance, if waves that have frequencies 500 HZ and 600 Hz are superposed, the variation of intensity (100 HZ) is too rapid for our ears to follow.

Audio 2.2. Frequencies 500 HZ and 600 Hz. Available at https://iopscience.iop.org/book/978-0-7503-3791-5.

We can also use the MATLAB program to check our ears! According to introductory textbooks, humans can hear sound in the range of 20 Hz to 20000 Hz. Using the MATLAB sound generator, we can check whether we can really hear this entire range (especially near the upper and lower limits).

Can you hear 80 Hz?

Audio 2.3. Frequency 80 HZ. Available at https://iopscience.iop.org/book/978-0-7503-3791-5.

You should be able to hear 1000 Hz and 300 Hz clearly.

Audio 2.4. Frequency 1000 HZ. Available at https://iopscience.iop.org/book/978-0-7503-3791-5.

Audio 2.5. Frequency 300 Hz. Available at https://iopscience.iop.org/book/978-0-7503-3791-5.

But what about 10 000 Hz or 12 000 Hz?

Audio 2.6. Frequency 10 000 Hz. Available at https://iopscience.iop.org/book/978-0-7503-3791-5.

Audio 2.7. Frequency 12 000 Hz. Available at https://iopscience.iop.org/book/978-0-7503-3791-5.

You might also notice that the apparent intensity of the sound in these audio files seems to vary with the frequency. This variation actually occurs because the response of the human ear depends on the frequency of the sound wave that is incident on it. The eardrum is forced to vibrate at the frequency of the external disturbance and the amplitude response is at a maximum at the resonant/natural frequency of the eardrum.

Programming exercises

2.1. Write a program that will decide whether a given number is prime or composite. The number is to be taken as input from the user (use the 'input' command). The program has to check whether the given number N is divisible by the numbers up to $\sqrt{N}$. (why?)

2.2. Write a program which will list all the prime numbers from 1 to 100. Use the 'sieve of Eratosthenes' to generate the list. The 'sieve of Eratosthenes' consists of writing down all the numbers (here, we store the numbers in a one-dimensional array) from two to 100. We know that two is a prime number. Every multiple of two is then struck off the list, since it is not a prime number. Similarly, all multiples of 3, 5, 7, ... etc. are struck off the list. When the program has run up to 100, the numbers that remain are the prime numbers.

2.3. Write a program which will generate the famous 'Pascal's triangle'. In case you have forgotten, Pascal's triangle is reproduced below:

1

1 2 1

1 3 3 1

1 4 6 4 1

1 5 10 10 5 1

1 6 15 20 15 6 1

1 7 21 35 35 21 7 1

Notice that each element in Pascal's triangle is equal to the sum of the elements to the left and right of the element in the previous row. The nth row of the triangle consists of the coefficients in the binomial expansion of $(a+b)^n$. You can use an array to store the triangle. Note that you will have zeros for the unused positions of the array (maybe you can avoid displaying the zeros).

For convenience, you may want to display the Pascal's triangle as follows:

```
1
1 2 1
1 3 3 1
1 4 6 4 1
```

```
1 5 10 10 5 1
1 6 15 20 15 6 1
1 7 21 35 35 21 7 1
```

This may be more convenient if you are using an array to store the elements of the triangle.

2.4. Write a program which will calculate the sine of a given angle (expressed in degrees). First, think of a procedure and then try to implement it. Hint: you can use a series expansion for the sine, but then the series can be used only if $x < 1$. How will you calculate sin x when the value of $x > 1$?

Chapter 3

Finding the roots and zeros of a function

3.1 The roots of a polynomial

You should be familiar with the procedure used to find the roots of a quadratic polynomial. While we initially learn to factorise the given polynomial, we soon memorise the formula that provides us with the roots.

The roots, α and β, of the polynomial ax^2+bx+c are given by:

$$\alpha = \frac{-b + \sqrt{b^2 - 4ac}}{2a} \qquad \beta = \frac{-b - \sqrt{b^2 - 4ac}}{2a} \qquad (3.1)$$

More cumbersome and complicated formulae have been derived for the roots of polynomials of degrees three and four. It is well known that similar formulae cannot be obtained for polynomials of degrees equal to or greater[1] than five.

Having realised that radicals cannot obtain the roots of quantic or higher-order polynomials, we are forced to resort to numerical methods to obtain the roots of these polynomials.

3.2 Graphical method

One of the simplest methods is to draw a graph of the given polynomial and locate the value of x for which the polynomial is zero. This method can be used to (approximately) obtain the zeros of any function $f(x)$, not just polynomials.

For example, consider the polynomial $x^3-7x^2-10x+16$. Even though its roots can be obtained by a formula (which I cannot remember!), we will determine its approximate roots by the graphical method. Note that even to obtain the approximate value of the roots, you need a vague idea of the range within which these roots occur! If you plot the polynomial over a very large range, then it may be difficult to locate the roots on

---

[1] This was shown by Evariste Galois (1811–1832), who died an untimely death in a duel. You can read about his life and work in *Men of Mathematics* by E T Bell, Simon and Schuster, New York, 1986. The topic of finding the roots of polynomials is also discussed in detail in *Why Beauty Is Truth* by Ian Stewart, Basic Books, New York, 2008.

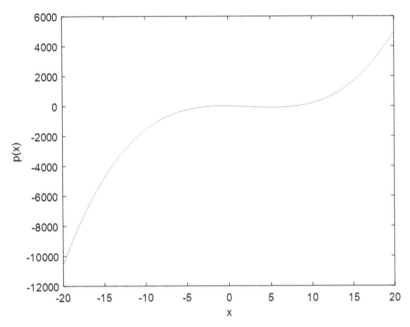

Figure 3.1. Plot of $x^3 - 7x^2 - 10x + 16$. It is clear that we have overestimated the range in which the roots lie.

the graph. It is clear that for very large positive values of x, the given polynomial will be positive (because the cubic term, which has a positive coefficient, will dominate). Similarly, for large negative values of x, the polynomial will have a negative value (again, because of the cubic term). Therefore, none of the roots can be too far from the origin. If the root is very close to the origin, the higher powers of x can be ignored (close to the origin), and an approximate root may be close to $x = 1.6$ (ignoring the quadratic and cubic terms). Considering the terms up to the quadratic term gives us 2.3864 and -0.9578 (as approximate roots). So let us see whether plotting the polynomial in the range $x = -20$ to $x = +20$ will give us a fair idea of its roots.

The MATLAB program used to obtain figure 3.1 is given below:
```
x=-20:0.1:20;
y=x.^3-7*x.^2-10*x+16;
plot(x,y)
```

The simple exercise above has taught us something invaluable. Unless we have a reasonable idea of the range over which the roots occur, a plot of the polynomial or function will not help us to get an approximate value for the roots. However, figure 3.1 is not entirely useless. We note from the figure that the roots are confined to the range $x = -10$ to $x = +10$. Let us now zoom into this range and see whether we can locate the roots from the graph (see figure 3.2). Placing the mouse pointer over a point in the MATLAB figure window will display the coordinates at that point.

Moving the mouse over the curve in the MATLAB figure window, we find that $x = 8$ is a root. Once we have obtained one root, α, we can divide the given

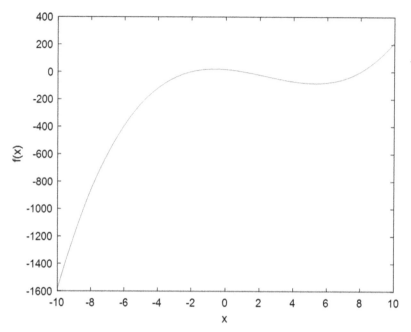

Figure 3.2. Plotting the same polynomial as in figure 3.1, but over a reduced range.

polynomial by $x-\alpha$ to get a quadratic polynomial whose roots can then be obtained with ease.

Dividing $x^3-7x^2-10x+16$ by $x-8$, or equivalently by eliminating the factor $x-8$, we get:

$$x^3 - 7x^2 - 10x + 16 = (x - 8)(x^2 + x - 2).$$

The roots of x^2+x-2 are -2 and $+1$.

The graphical method can also be used to obtain approximate solutions for equations involving trigonometric or other functions, as the following example shows.

Example 3.1

The intensity of the Fraunhofer diffraction pattern[2] of a single slit of width b is given by:

$$I = I_o \frac{\sin^2(kx)}{(kx)^2} \tag{3.2}$$

Here, $x = (b \sin\theta)/2$ and $k = 2\pi/\lambda$. Determine the location of the maxima of the intensity pattern.

Answer: the maxima of the diffraction pattern are obtained by differentiating (3–2) with respect to x and setting the derivative to be equal to 0.

---

[2] See, for instance, *Optics* by Ajoy Ghatak, 2nd edn, Tata McGraw Hill, New Delhi, 2001.

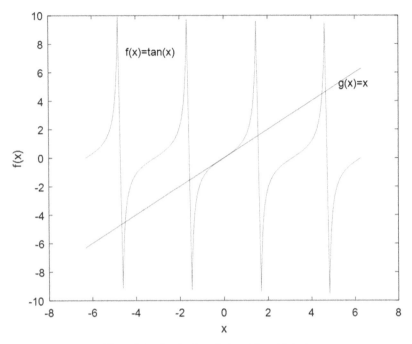

Figure 3.3. Graph of $\tan(kx)$ and kx with $k=1$.

$$\frac{dI}{dx} = \frac{2I_o \sin(kx)k \cos(kx)}{(kx)^2} - \frac{2I_o \sin^2(kx)}{k^2 x^3} = 0 \qquad (3.3)$$

$$\frac{dI}{dx} = \frac{2I_o \sin(kx)}{kx^2}\left[\cos(kx) - \frac{\sin(kx)}{kx}\right] = 0 \qquad (3.4)$$

The solutions to equation (3.4) are:

$x \to \infty$, $\sin(kx) = 0$ (which corresponds to the minima), and $kx = \tan(kx)$

To locate the maxima of the diffraction pattern, we can either locate the zeros of $\tan(kx) - kx$ or plot kx and $\tan(kx)$ separately and locate the points where they are equal. Plotting the two functions separately is easier, as we already have an intuitive feel for the behaviours of these two functions.

Figure 3.3 shows the plots for $\tan(kx)$ and kx with $k = 1$. In drawing these graphs, we have excluded those points where $\tan(x)$ has a very large value. It can be seen from the figure that given the nature of $\tan(x)$, there will be an infinite number of points at which $\tan(x)$ is equal to x. We see that one of the solutions must lie between $x = \pi$ and $x = 3\pi/2$. Note that the straight line appears to cut the graph of $\tan x$ at $\pi/2$, but remember that $\tan x$ is discontinuous at $\pi/2$. Given the rapid variation in $\tan(x)$, we cannot locate the solution with reasonable accuracy. We will need better numerical methods to determine the roots/solutions accurately. We now start our journey to learn accurate techniques for determining the roots or zeros of a function.

3.3 Solution of equations by fixed-point iteration

In the fixed-point iteration method, we rewrite the given equation so that it has the form

$x = f(x)$. We start with an initial guess for the solution to this equation, say, x_o. Substituting this initial guess into the given equation gives us:

$x_1 = f(x_o)$. In general, x_1 will not be equal to x_0, unless it is the desired solution. We now iterate this equation until it converges to the actual solution for x; i.e. we determine $x_2 = f(x_1)$, $x_3 = f(x_2)$ or, in general:

$x_{i+1} = f(x_i)$. Each iteration is supposed to take us closer and closer to the actual solution of this equation. But, what is the guarantee that this procedure converges—in other words, will we move away from the solution, instead of approaching it?

Let us define the error (ε_i) at a particular iteration i to be the difference between the actual solution x and the approximation to the solution at that iteration, i.e. x_i.

$$\varepsilon_{i+1} = x - x_{i+1}$$

$$\varepsilon_i = x - x_i$$

Subtracting x from both sides of the equation $x_{i+1} = f(x_i)$:

$x_{i+1} - x = f(x_i) - x$

$-\varepsilon_{i+1} = f(x) + (x_i - x)f'(x) - x$,

where we have performed a Taylor series expansion of $f(x_i)$ *about* x. The equation above can be written as:

$-\varepsilon_{i+1} = - -\varepsilon_i f'(x)$ since $f(x) = x$

To converge the iterative procedure, we require $|\varepsilon_{i+1}| < |\varepsilon_i|$; therefore, the procedure will converge if $f'(x) < 1$. We are trying to determine the solution for x. However, since x_i is close to x, we can check for convergence by verifying that $f'(x_i) < 1$.

Example 3.2

Verify that the iterative procedure described above can be used to determine the roots of the polynomial $x^3-7x^2-10x+16$. Determine the roots if the method converges.

Answer: we are asked to determine the solutions to the equation $x^3-7x^2-10x+16 = 0$. To use the fixed-point iteration method, we need to rewrite this in the form $x = f(x)$. The given equation can be rewritten as:

$$x = \frac{x^3 - 7x^2 + 16}{10}. \tag{3.5}$$

The derivative of $f(x)$ is found to be:

$$f'(x) = \frac{3x^2 - 14x}{10}. \tag{3.6}$$

We see that for all three roots, i.e. eight, one, and minus two, the magnitude of the derivative is greater than one. Therefore, the iterative method cannot be used to find the roots if (3.5) is used.

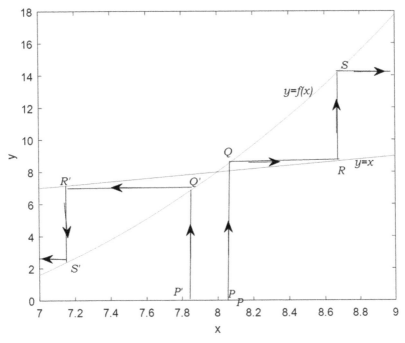

Figure 3.4. Graphical representation of the divergence of successive values of x if the iteration method is used on (3.5).

The results of using the iterative method on (3.5) are shown in figure 3.4. It can be seen from the figure that starting with a value of x slightly greater than eight (i.e. point P), when substituted into $f(x)$ (i.e. into (3.5)), gives a value much greater than the starting value of the next iteration. The iterative method is shown graphically in figure 3.4. We draw a vertical line from the initial guess (P) for the root to the point where this vertical line intersects the curve $f(x)$, point Q. Since we are using the equation $x = f(x)$, we draw a horizontal line from Q until it intersects the straight line $y = x$. The point of intersection (R) will be the value of x for the next iteration. The successive iterations can be produced graphically by alternating between horizontal and vertical lines. It is clear from the figure that starting with a value slightly greater than eight leads to larger and larger values of x and that the process does not converge to eight. Similarly, starting with a value slightly less than eight (i.e. point P') produces the points Q', R', and S' successively, again leading us away from eight, which is the correct root. Of course, if we start with exactly $x = 8$, we will remain at eight, but then, if the root is unknown, the probability of accidentally choosing the correct root is almost zero.

It should be noted that there are other ways of rewriting $x^3-7x^2-10x+16 = 0$ in the form $x = f(x)$. Suppose we rewrite $x^3-7x^2-10x+16 = 0$ as:

$$x = \pm\sqrt{\frac{x^3 - 10x + 16}{7}}. \tag{3.7}$$

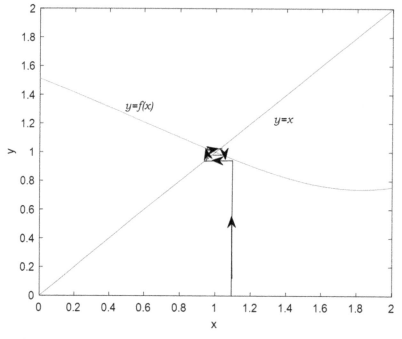

Figure 3.5. If the initial value chosen is close to $x = 1$, the value converges to $x = 1$ after a few iterations.

The derivative of the right-hand side of (3.7) is:

$$f'(x) = \pm \frac{1}{2} \frac{(3x^2 - 10)}{\sqrt{7(x^3 - 10x + 16)}}.$$
(3.8)

It can be seen that the condition for convergence is now satisfied for $x = 1$ and $x = -2$, but not for $x = 8$.

Figure 3.5 graphically shows the convergence to the root at $x = 1$. If the initial value is close to $x = 1$ (say 1.1), then the value converges to one after a few iterations.

Therefore, the convergence of the iterative procedure depends on how we rewrite the given equation in the form $x = f(x)$. We would like a procedure that converges unconditionally. Therefore, the fixed-point iteration method is of limited use in determining the solutions of algebraic equations. We now discuss the bisection method, which converges for almost all possible equations.

3.4 Bisection

In the bisection method, we are looking for the point x at which $f(x) = 0$. The bisection method uses the fact if $f(a) > 0$ and $f(b) < 0$, then there must be a point x (which lies in the interval (a,b)), such that $f(x) = 0$. In the next iteration, we calculate $c = (a+b)/2$, i.e. we bisect the interval (a,b) and determine $f(c)$. If $f(c)$ is positive, then the root x is between b and c; otherwise, it is between a and c. In a computer program that applies the bisection method, if the root is between b and c, then c

replaces a in the next iteration, while if the root is between a and c, c replaces b in the next iteration.

Let us use this method to find solutions to the equation $\tan x = x$ (example 3.1).

As we mentioned earlier, one solution to this equation lies between π and $3\pi/2$. Therefore, we use these as starting values for the bisection method.

To use the bisection method, we rewrite the given equation as $\tan x - x = 0$ and note that $\tan\pi-\pi$ is negative, while $\tan(3\pi/2)-3\pi/2$ is positive (we approach $3\pi/2$ from the lower side). Using these two values as starting values for the bisection method, we have $c = 5\pi/4$. We find that $\tan(5\pi/4)-5\pi/4$ is negative. Therefore, the solution is between $5\pi/4$ and π. Repeated application of the bisection method gives us the solution as 4.492 (accurate to the third decimal place).

```
'Program for bisection method';
a=3.14;
b=1.5*3.14;
tol=5e-3;
error=abs(b-a);
fa=FofX3(a);
fb=FofX3(b);
iterations=0;
while (error>tol)
    c=(a+b)/2;
    fc=FofX3(c);
    iterations=iterations+1;
    if(fa*fc)<=0
        b=c;
        fb=fc;
    else
        a=c;
        fa=fc;
    end
    error=abs(b-a);

end
disp(c)
    disp(iterations)
```

Note that the procedure is iterated until the range (a,b) drops below the tolerance specified by the variable tol. When the error (i.e. $|b-a|$) is smaller than tol, the while loop is terminated. This is a general feature of any iterative procedure. We can never look for the exact equality of a quantity to zero in order to stop an iterative procedure. Due to rounding errors, exact equality can never be achieved. In addition, note that in the 'if' condition used in the program, we have used the product of the function at $x = a$ and $x = c$. This product will be negative if $f(a)$ and $f(c)$ have opposite signs. If the product is negative, then the root lies between a and c, otherwise it lies between b and c. This way of checking the interval within which the root lies is more efficient than using multiple 'if' conditions to check the location of the root.

Can we use the bisection method to find the roots of the polynomial $x^8+x^6+x^4+x^2+1$? After a little thought, you will realise that this polynomial is positive for all values of x (since it contains only even powers of x and the constant term is also positive). As a result, we do not have two points at which the function has opposite signs, which is the basic premise of the bisection method. All the roots of this polynomial are complex. As the bisection method bisects an interval on a real line, it will not help us to locate roots which lie elsewhere on the complex plane. However, we can tweak this method to determine the complex roots. We start by rewriting the given polynomial as $z^8+z^6+z^4+z^2+1$ and use the fact that $z = R\exp(i\theta)$ to obtain the following equations.

$$R^8\cos8\theta + R^6\cos6\theta + R^4\cos4\theta + R^2\cos2\theta + 1 = 0 \qquad (3.9a)$$

$$R^8\sin8\theta + R^6\sin6\theta + R^4\sin4\theta + R^2\sin2\theta = 0. \qquad (3.9b)$$

Since we have two equations with two unknowns, let us eliminate one of the variables to obtain a single equation to which we can apply the bisection method. The second equation gives us: $R^2(R^6\sin8\theta + R^4\sin6\theta + R^2\sin4\theta+\sin2\theta) = 0$. Since $R = 0$ is not consistent with the original equation, we take the bracketed term and rewrite it as:

$$R^2 = -\frac{(R^6\sin8\theta + R^4\sin6\theta + \sin2\theta)}{\sin4\theta}. \qquad (3.10)$$

Substituting for R^2 in equation (3.9a), we obtain an equation containing only θ, to which we can apply the bisection method. However, if you try this by hand, you will soon realise that this is a cumbersome procedure; hence, the bisection method is not used to determine the complex roots of a polynomial.

The above discussion tells us that we need an easy way of determining whether a given polynomial has complex roots. The Descartes rule of signs can be used to give us an indication of the presence of complex roots.

3.5 Descartes' rule of signs

Descartes' rule of signs states that the number of real positive roots of a polynomial can, at most, be equal to the number of times the signs of the coefficients change in a polynomial. If complex roots occur, then the number of positive roots can be less than the number of sign changes by an even number (since complex roots occur in conjugate pairs). Hence, Descartes' rule can be used to obtain an upper bound on the number of real positive roots.

Similarly, there is an upper bound on the number of real negative roots. To obtain the upper bound on the number of real negative roots, replace x by $-x$ in the given polynomial and count the number of times the signs of the coefficients change. In the polynomial discussed earlier, i.e. $x^8+x^6+x^4+x^2+1$, we observe no changes in sign, so the number of positive real roots is zero. Replacing x with $-x$ does not lead to any change in the polynomial, so the number of negative real roots is also zero.

Example 3.3

Apply Descartes' rules to determine the number of positive real roots and the number of negative real roots for the polynomial $x^3-3x^2+4x-6 = 0$.

Answer: we note that there are three changes of sign in the given polynomial. As a result, there can be a maximum of three positive real roots. Replacing x with $-x$ leads to a polynomial in which all terms have a negative sign. Therefore, the given polynomial has no negative roots.

Note that Descartes' rule helps to narrow down the range within which the roots can occur. We note that the given polynomial is negative at $x = 0$ and positive at $x = 3$ (the cubic term will dominate for large positive x; therefore, the polynomial will be positive for sufficiently large values of x). Using these as starting values, we obtain 2.376 as one of the roots.

We now summarise the drawbacks of the bisection method:

   (i) It requires two starting values.

  (ii) Since the interval is halved in each iteration, the convergence is rather slow (it is linear).

 (iii) It is not suitable for the determination of complex roots.

The only advantage of the bisection method is that it is guaranteed to converge to the solution if suitable starting values are chosen.

We now discuss the Newton–Raphson method, which overcomes all the drawbacks of the bisection method.

3.6 The Newton–Raphson method

As in the bisection method, in the Newton–Raphson method, we are looking for the point x at which $f(x) = 0$.

The Newton–Raphson method approximates the curve for which we are trying to find the zeros by a straight line. The straight line is the tangent to the curve at the point which is chosen as the initial approximation to the root. The next estimate of the root is the point at which the tangent intersects the x-axis. We can derive the relation between the estimates x_i and x_{i+1} in successive iterations by using the Taylor series.

We use the Taylor series to expand the function around x_i. Let x be the actual root, then:

$$f(x) = f(x_i) + (x - x_i)f'(x_i). \tag{3.11}$$

Using the fact that $f(x) = 0$, we can rewrite this equation as:

$$x = x_i - \frac{f(x_i)}{f'(x_i)}. \tag{3.12}$$

Even though we started by saying that x is the actual root, the value of x obtained using (3.12) is only a better approximation to the actual root (when compared with x_i). This is because we ignored higher-order terms in the Taylor series to obtain (3.11).

To obtain better and better approximations to the root, we have to iterate (3.12); therefore, we rewrite it as follows:

$$x_{i+1} = x_i - \frac{f(x_i)}{f'(x_i)}. \tag{3.13}$$

Note the advantages of the Newton–Raphson method over the bisection method. We require only one starting value instead of the two needed for the bisection method. Also, we can show that its convergence is faster than that of the bisection method. Before we look at the rate of convergence of the Newton–Raphson method, let us examine how it converges to the root (graphically).

We will try to locate the roots of $f(x) = x^3 - 7x^2 - 10x + 16$ using the Newton–Raphson method. Figure 3.6 shows the graph of this polynomial close to $x = 8$. If we choose $x = 9$ as the starting value for the Newton–Raphson method, we can then draw a tangent to the curve at $x = 9$. From figure 3.6, we see that the tangent intersects the x-axis at approximately $x = 8.15$. Therefore, in the next iteration of the method, we draw a tangent to the curve at $x = 8.15$. It can be seen that the convergence to $x = 8$ is quite fast.

We now discuss the algebraic implementation of the Newton–Raphson method. To apply the Newton–Raphson method to the given polynomial, we differentiate the

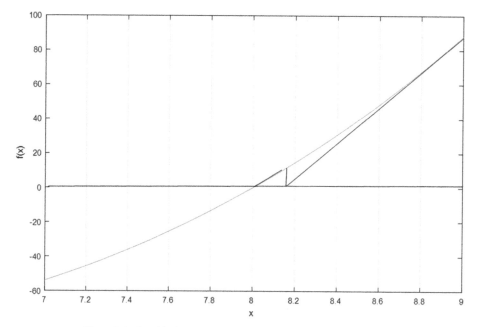

Figure 3.6. Graphical representation of the Newton–Raphson method.

given polynomial with respect to x and substitute the polynomial and its derivative into (3.13). This gives us:

$$x_{i+1} = x_i - \frac{(x^3 - 7x^2 - 10x + 16)}{(3x^2 - 14x - 10)}.$$ (3.14)

To determine the rate of convergence, we should determine the rate at which the error (the difference between the approximate root and the actual root) decreases in successive iterations. As in section 3.3, we define the error in successive iterations as:

$$\varepsilon_{i+1} = x - x_{i+1}$$
$$\varepsilon_i = x - x_i.$$ (3.15)

Here, x is the actual root and x_{i+1} and x_i are approximations to the root in the $(i+1)$th and ith iterations, respectively.

Subtracting x from both sides of (3.13), we have:

$$x_{i+1} - x = x_i - x - \frac{f(x_i)}{f'(x_i)}.$$ (3.16)

Using (3.15) and the Taylor series expansion of $f(x_i)$ about x, (3.16) can be rewritten as:

$$-\varepsilon_{i+1} = -\varepsilon_i - \frac{\left(f(x) - \varepsilon_i f'(x_i) - \frac{\varepsilon_i^2}{2} f''(x_i) \right)}{f'(x_i)}.$$ (3.17)

Equation (3.17) simplifies to:

$$-\varepsilon_{i+1} = -\frac{\left(\frac{\varepsilon_i^2}{2} f''(x_i) \right)}{f'(x_i)}.$$ (3.18)

From (3.18), we see that the error at one iteration is proportional to the square in the previous iterations. Since the error is supposed to be small, it reduces quadratically. This is much faster than the linear rate of convergence achieved by the bisection method.

The fact the Newton–Raphson method converges quadratically may make it seem that it is far superior to the bisection method, which converges linearly, but there is a problem lurking within the Newton–Raphson method: it may not converge at all in some cases!

This problem is evident if we study (3.13) more closely. We see that the derivative of the function is in the denominator of the second term of the right-hand side. What happens if the derivative is zero at the point chosen to be the initial value of x? The value of x in the next iteration tends to infinity. We can also visualise this graphically. If the derivative at a point is zero, then the tangent to the curve is a horizontal line that never intersects the x-axis. In fact, in practice, the

Newton–Raphson method may not converge even if the derivative is nonzero but very small. The reciprocal of this small number may be so large that it cannot be stored on your computer. Therefore, the convergence of the Newton–Raphson method should be verified before we use it.

We might want to design a hybrid method that combines the bisection method's reliability with the rate of convergence of the Newton–Raphson method. It is possible to have the best of both methods. Assume that we have chosen two points $x = a$ and $x = b$, such that $f(a).f(b) < 0$, i.e. the actual root lies between a and b. We now use $c = (a+b)/2$ as a starting value for the Newton–Raphson method. All subsequent iterations are performed using the Newton–Raphson method as long as the values of x lie between a and b. When we see that the value of x lies outside the interval (a,b), we switch back to the bisection method.

Example 3.4

Apply the Newton–Raphson method to determine the square root of a number N. Apply this method to determine the square root of 40.

Answer: let x be the square root of N. We are now trying to determine the solution to the equation $x^2 = N$ or $x^2 - N = 0$.

Therefore, in this case, $f(x) = x^2 - N$ and $f'(x) = 2x$. Substituting these into (3.13), we have:

$$x_{i+1} = x_i - \frac{(x_i^2 - N)}{(2x_i)} = \frac{1}{2}\left(x_i + \frac{N}{x_i}\right). \tag{3.19}$$

We now iterate (3.19) to determine the square root of N. If $N = 40$, let us take $x = 6$ as our initial guess for the square root. The value of x in the next iteration will be $0.5(6 + 40/6) = 6.3333$. The subsequent values are 6.324561, 6.324555, 6.324555.... Therefore, we have obtained the square root accurate to the sixth decimal place in just four iterations.

3.7 The false position method

This is a variation on the bisection method. In the bisection method, the root is bracketed between $x = a$ and $x = b$. Suppose we join the points $(a, f(a))$ and $(b, f(b))$ by a straight line; the point of intersection of this line with the x-axis, i.e., $x = c$, will then be the next approximation to the root. Then, just as in the bisection method, a or b will be replaced by c in the next iteration.

The equation for the line that passes through the points $(a, f(a))$ and $(b, f(b))$ is:

$$\frac{y - f(a)}{x - a} = \frac{f(b) - f(a)}{b - a}. \tag{3.20}$$

In the above equation, we let $x = c$ and $y = 0$ to get the next approximation to the root.

$$x = a - \frac{(b - a)f(a)}{f(b) - f(a)}. \tag{3.21}$$

In most cases, the false position method converges faster than the bisection method, but in some instances, it can also converge at a rate slower than that of the bisection method.

3.8 The secant method

The secant method is a variation of the Newton–Raphson method. We approximate the derivative of the function using the values of the function at x_i and x_{i-1}, which are the two previous approximations to the root.

$$f'(x) = \frac{f(x_i) - f(x_{i-1})}{x_i - x_{i-1}} \qquad (3.22)$$

Substituting this expression[3] for the derivative into (3.13) gives us:

$$x_{i+1} = x_i - f(x_i)\frac{(x_i - x_{i-1})}{f(x_i) - f(x_{i-1})}. \qquad (3.23)$$

While (3.23) looks very similar to (3.21), in (3.21), a and b are two points at which the function has opposite signs, whereas x_i and x_{i-1} are two successive approximations to the root, and the function can even have the same sign at these two points.

3.9 Applications of root finding in physics

Finding the roots of a polynomial or the zeros of a function has a wide variety of applications in many areas of physics. We give a few examples to illustrate the application of the root-finding methods we have discussed.

Example 3.5
The energy density within an isothermal blackbody is given by Planck's radiation law:

$$u(\lambda)d\lambda = \frac{8\pi hc}{\lambda^5} \cdot \frac{1}{\exp\left(\dfrac{hc}{\lambda kT}\right) - 1}.$$

This has a maximum at a certain wavelength, $\lambda_{\max}$. Use the expression for the energy density given above to derive an equation which can be used to determine $\lambda_{\max}$. This leads to Wien's displacement law, which states that $\lambda_{\max}T = \frac{hc}{k}*w$, where w is a numerical constant. Use the Newton–Raphson method to determine the value of w.

Answer: The maximum in the blackbody spectrum can be located by differentiating the energy density u with respect to the wavelength and setting the derivative equal to zero:

---

[3] Equation (3.22) is a numerical approximation to the derivative at x_i, known as the backward difference formula. We will encounter it in the discussion of numerical differentiation.

$$\frac{du}{d\lambda} = \frac{-5 \times 8\pi hc}{\lambda^6} \times \frac{1}{\left[\exp\left(\dfrac{hc}{\lambda kT}\right) - 1\right]} + \frac{8\pi hc}{\lambda^7} \times \frac{\exp\left(\dfrac{hc}{\lambda kT}\right)}{\left[\exp\left(\dfrac{hc}{\lambda kT}\right) - 1\right]^2} \times \frac{hc}{\lambda^2 kT} = 0.$$

Letting $x = \dfrac{hc}{\lambda kT}$ and ignoring the solution $1/\lambda = 0$, which occurs at infinity, we get the equation:

```
xexp(x)=5(exp(x)-1)
Or  f(x)=x+5exp(-x)-5=0
fa'(x)=1-5exp(-x)
```

Choose `x=2` as the starting value and substitute it into the equation (for the Newton–Raphson method):

$$x_1 = x_o - \frac{f(x)}{f'(x)} = 1 - \frac{5\exp(-1) - 4}{1 - 5\exp(-1)}.$$

The successive values of x are 9.18 and 4.99 (the value converges to 4.965 after a few iterations). Note that taking a value close to one or zero will cause the method to converge to the solution at $x = 0$, which is a physically meaningless solution.

$$w = 1/x \sim 0.2$$

Planck's explanation of the blackbody spectrum led to the birth of Quantum mechanics. Therefore, it is appropriate for us to consider another application of root finding from Quantum mechanics.

3.10 The finite potential well

One of the first problems in a standard course on quantum mechanics is the determination of the energy eigenvalues of an infinite potential well. We will quickly recap this problem before we go on to the more realistic finite potential well.

Consider a particle of mass m confined to a one-dimensional infinite potential well extending from $x = -a$ to $x = +a$. The particle can move freely between these limits, as the potential is zero, but the potential changes to infinity at the boundaries of the well.

The wave function describing the state of such a system satisfies the Schrödinger equation[4]:

$$-\frac{\hbar^2}{2m}\frac{d^2\psi}{dx^2} + V(x)\psi(x) = E\psi(x). \tag{3.24}$$

---

[4] An excellent introduction to the fundamentals of Quantum Physics is provided in *Quantum Physics of Atoms, Molecules, Solids, Nuclei, and Particles* by Robert Eisberg and Robert Resnick, 2nd edn, Wiley India, New Delhi, 2009.

Since, within the region, the potential is zero, (3.24) can be rewritten as:

$$\frac{d^2\psi}{dx^2} + \frac{2mE}{\hbar^2}\psi(x) = 0. \tag{3.25}$$

The solutions to this equation are clearly oscillatory, and we can also use the boundary condition that the wave function should vanish at the boundaries of the well (using the fact that the wave function is zero outside and invoking the continuity[5] of the wave function).

The solution of (3.25) can be written as:

$$\psi(x) = A \cos\left(\frac{\sqrt{2mE}}{\hbar}x\right), \tag{3.26}$$

where A is a normalisation constant. Invoking the fact that the wave function vanishes at the boundaries, we have:

$$\frac{\sqrt{2mE}}{\hbar} = \frac{(2n+1)\pi}{2a} \tag{3.27}$$

or

$$E_n = \frac{(2n+1)^2\pi^2\hbar^2}{8ma^2}. \tag{3.28}$$

The other possible solution is:

$$\psi(x) = A \sin\left(\frac{\sqrt{2mE}}{\hbar}x\right). \tag{3.29}$$

Invoking the boundary condition on the wave function, we have:

$$\frac{\sqrt{2mE}}{\hbar} = \frac{n\pi}{a} \tag{3.30}$$

or

$$E_n = \frac{n^2\pi^2\hbar^2}{2ma^2}. \tag{3.31}$$

Such a straightforward (analytical) evaluation of the energy eigenvalues and the corresponding wave functions is not possible if the potential well has a finite height. In the case of a finite potential well, the wave function need not change to zero at the boundaries of the well. However, we can still use the boundary conditions on the wave function and its first derivative to arrive at an equation which can be solved using numerical methods.

---

[5] If the wave function is discontinuous then the probability density would change discontinuously, which is nonphysical.

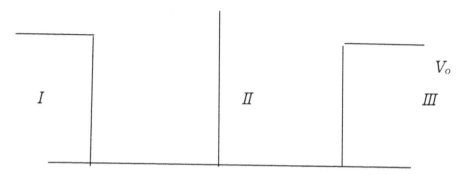

Figure 3.7. A finite potential well.

Consider a finite potential well of height V_o shown in the figure given below (figure 3.7).

For the finite potential, in region *II*, the Schrödinger equation is similar (3.25) to that of the infinite well, whereas for regions *I* and *III*, the equation now reads:

$$\frac{d^2\psi}{dx^2} - \frac{2m(V_o - E)}{\hbar^2}\psi(x) = 0. \tag{3.32}$$

In region *I*, the solution to this equation is of the form:

$$\psi_I = C\exp(\beta x). \tag{3.33}$$

Note that the wave function has to decay exponentially outside the well so that it goes to zero at infinity. Since x is negative in region *I*, the solution $D\exp(-\beta x)$ is not allowed, as it would lead to a solution which increases exponentially with x.

In equation (3.33), $\beta = \sqrt{\dfrac{2m(V_o - E)}{\hbar^2}}$.

In region *II*, the solution is of the form:

$$\psi_{II} = A\sin(\alpha x) + B\cos(\alpha x). \tag{3.34}$$

Note that we have to include both sine and cosine terms in the general solution, as the wave function need not vanish at the boundaries.

Where $\alpha = \sqrt{\dfrac{2mE}{\hbar^2}}$,

using the fact that the wave function should be continuous across the boundary of the well at $x = -a$, we have:

$$C\exp(-\beta a) = -A\sin(\alpha a) + B\cos(\alpha a). \tag{3.35}$$

Using the fact that the derivative of the wave function has to be continuous[6] at the same point:

---

[6] If the derivative of the wave function is discontinuous at a point then the second derivative will become infinite at that point, which corresponds to the kinetic energy of the particle becoming infinite.

$$C\beta \exp(-\beta a) = A\alpha \cos(\alpha a) + B\alpha \sin(\alpha a). \tag{3.36}$$

Applying the continuity of the wave function and its derivative at $x = a$, we have:

$$D \exp(-\beta a) = A \sin(\alpha a) + B \cos(\alpha a) \tag{3.37}$$

$$-\beta D \exp(-\beta a) = A\alpha \cos(\alpha a) - B\alpha \sin(\alpha a). \tag{3.38}$$

Since the potential is symmetric about the origin, we expect the wave function to have a definite parity. Therefore, $A = 0$ or $B = 0$. If we let $A = 0$ and divide (3.36) by (3.35), we get:

$$\beta = \alpha \tan(\alpha a). \tag{3.39}$$

If we let $B = 0$ and divide (3.36) by (3.35), we get:

$$\beta = -\alpha \cot(\alpha a). \tag{3.40}$$

Note that these are a pair of transcendental equations and can only be solved numerically. Before attempting to solve these equations numerically, we should note that the constants include the Planck constant, which has a very small value in SI units. Therefore, it is more appropriate to use a different set of units in order to minimise rounding errors.

It is appropriate to use nanometers for the width of the well, the mass of the particle can be stated in terms of the electronic mass and the height of the potential well can be expressed in electron volts. We can now work out the magnitude of $\hbar^2$ in this system of units.

$$\text{We have} \quad \hbar^2 = \frac{h^2}{4\pi^2} = \frac{1}{4\pi^2}\left[6.6 \times 10^{-34} \text{J s}^{-1}\right]^2. \tag{3.41}$$

We can convert J into eV in (3.41), but we also need to convert s into this new system of units. To convert 1 s into the new system of units, we observe that $1 \text{ J} = 1 \text{ kg m}^2 \text{ s}^{-2}$.

$$\text{Or 1s} = \sqrt{\frac{\text{kgm}^2}{\text{J}}} = \sqrt{\frac{m_e}{9.1 \times 10^{-31}} \times \frac{10^{18}\text{nm}^2 \times 1.609 \times 10^{-19}}{1 \text{ eV}}}. \tag{3.42}$$

Substituting for 1 s from (3.42) into (3.41), we get

$$\hbar^2 = \frac{h^2}{4\pi^2} = \frac{1}{4\pi^2}[6.6 \times 10^{-34} \text{ J s}^{-1}]^2 = 0.076199682 m_e \text{ eV nm}^2.$$

Using this system of units, we find (for example) that for a finite square well of height 10 eV and width 0.3 nm, the following are the energy eigenvalues for the first two even eigenstates: 0.7155 eV and 6.1486 eV. The first odd eigenstates are found to be at 2.8214 eV and 9.8672 eV.

How do these energy eigenvalues compare with those for an infinite potential well? The first two energy eigenvalues for the even states of an infinite potential well

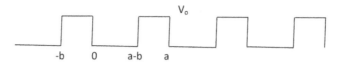

Figure 3.8. A periodic array of potential wells/barriers.

are: 1.0445 eV and 9.4008 eV, while the first two odd states are found to be at 4.1781 eV and 16.7125 eV.

The energy eigenvalues are smaller for a finite well, as it is effectively like having a wider infinite square well (the particle leaking outside corresponds to a larger well).

3.11 The Kronig–Penney model

The Kronig–Penney model is used in solid-state physics[7] to explain the band structures of solids.

We assume a periodic array of wells separated by a distance a. We also assume periodic boundary conditions. These two statements lead to the two following equations:

$$V(x + a) = V(x) \tag{3.43a}$$

$$\Psi(x + Na) = \Psi(x). \tag{3.43b}$$

Equation (3.43a) refers to the fact that the potential is periodic with a period a and the wave function at the Nth well is the same as the wave function at the first well. Note that the wave function does not have the periodicity[8] a (figure 3.8).

From Bloch's theorem, we have $\Psi(x+a) = A\Psi(x)$:

$$\Psi(x + 2a) = A^2 \Psi(x)$$
$$\ldots \ldots \Psi(x + Na) = A^N \Psi(x).$$

and from the periodic boundary condition (3.43b), we have $A^N = 1$, which gives $A = \exp\left(\dfrac{2\pi i m}{N}\right)$, where m is an integer. The Schrödinger equation for $0 \leqslant x \leqslant a-b$ is:

$$\frac{d^2\psi}{dx^2} + \frac{2mE\psi}{\hbar^2} = 0. \tag{3.44a}$$

The Schrödinger equation for $-b \leqslant x \leqslant 0$ is:

$$\frac{d^2\psi}{dx^2} + \frac{2m(E - V_o)\psi}{\hbar^2} = 0. \tag{3.44b}$$

[7] *Introduction to Solid State Physics*, Charles Kittel, 8th edn, Wiley, New York, 2012.
[8] This is obtained from Bloch's theorem. See *Introduction to Solid State Physics*, Charles Kittel, 8th edn, Wiley, New York, 2012 for more details.

The solution to (3.44a) is

$$\psi(x) = A \exp(i\alpha x) + B \exp(-i\alpha x), \tag{3.45a}$$

and the solution to (3.44b) is

$$\psi(x) = C \exp(\beta x) + D \exp(-\beta x). \tag{3.45b}$$

Here, $\beta = \sqrt{\dfrac{2m(V_o - E)}{\hbar^2}}$ and $\alpha = \sqrt{\dfrac{2mE}{\hbar^2}}$.

The wave function has to be continuous at all points. Using the fact that it is continuous at $x = 0$, we have:

$$A + B = C + D \tag{3.46}$$

The derivative should also be continuous at $x = 0$. Invoking this condition leads to:

$$i\alpha A - i\alpha B = C\beta - \beta D. \tag{3.47}$$

We now use Bloch's boundary condition, i.e.:

$$\psi(x + a) = \exp(ika)\psi(a) \tag{3.48}$$

at $x = -b$, to get:

$$\begin{aligned} & A \exp(i\alpha(a - b)) + B \exp(-i\alpha(a - b)) \\ & = \exp(ika)[C \exp(-\beta b) + D \exp(\beta b)]. \end{aligned} \tag{3.49}$$

Using (3.48) for the first derivative, we have:

$$\begin{aligned} & i\alpha A \exp(i\alpha(a - b)) - i\alpha B \exp(-i\alpha(a - b)) \\ & = \exp(ika)[\beta C \exp(-\beta b) - \beta D \exp(\beta b)]. \end{aligned} \tag{3.50}$$

For equations (3.46), (3.47), (3.49), and (3.50), we have a nontrivial solution only if the following determinant vanishes:

$$\begin{vmatrix} 1 & 1 & 1 & 1 \\ i\alpha & -i\alpha & -\beta & \beta \\ \exp(i\alpha(a - b)) & \exp(-i\alpha(a - b)) & -\exp(ika)\exp(-\beta b) & -\exp(ika)\exp(\beta b) \\ i\alpha \exp(i\alpha(a - b)) & -i\alpha \exp(-i\alpha(a - b)) & -\beta \exp(ika)\exp(-\beta b) & \beta \exp(ika)\exp(\beta b) \end{vmatrix} = 0.$$

By expanding this determinant and simplifying, we arrive at the equation:

$$\cos ka = \frac{(\beta^2 - \alpha^2)}{\alpha\beta} \sinh(\beta b)\sin[\alpha(a - b)] + \cosh(\beta b)\cos[\alpha(a - b)]. \tag{3.51}$$

Figure 3.9 shows a plot of the right-hand side of (3.51) for different values of the energy (in eV). Here, we have followed the same system of units that we adopted in section 3.10 for the finite potential well, using $a = 1$ nm, $b = 0.1$ nm, and $V_0 = 10$ eV. Since the left-hand side of (3.51) can only take values between $+1$ and -1, the energy

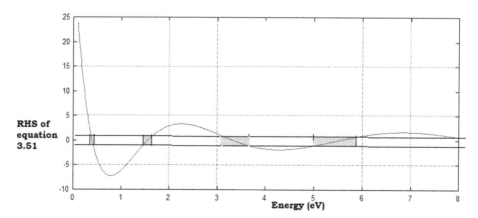

Figure 3.9. Plot of the right-hand side of equation 3.51 for various energy values. The shaded region corresponds to the allowed energy bands. We have chosen $a = 1$ nm, $b = 0.1$ nm, and $V_0 = 10$ eV.

values for which the right-hand side is greater than one correspond to physically unrealisable or forbidden energies.

It can be seen from figure 3.9 that some continuous ranges of energy values are allowed (shown shaded in the figure). The unshaded ranges of energy values are the energy gaps that are forbidden. Therefore, the Kronig–Penney model can provide a theoretical justification for the experimentally determined band structure in solids. Note that the bands get 'stretched' as we move towards higher energy values. It is to be expected that the energy spectrum does not have forbidden gaps beyond 10 eV, as the height of the barrier is 10 eV in this case.

Exercises

3.1. How many real solutions does the equation $\sin x = \dfrac{x}{10\pi}$ have? Over what range of x do these solutions occur? Draw a graph of the left-hand side and right-hand side of this equation to estimate the solutions. Write a program that uses the bisection method to obtain all the solutions. The program should automatically choose different starting values to obtain all the solutions one after the other.

3.2. How many roots does the equation $\cos x - x = 0$ have? Drawing graphs of $\cos(x)$ and x will help you to estimate the roots.

3.3. How many zeros does the function $f(x) = x^4 + \dfrac{\cos(40x)}{40}$ have? Over what range of x do these solutions occur?

3.4. A 1 kg mass on a frictionless table is attached to one end of a massless spring with a natural length of 1 m and a spring constant $k = 1$ N m^{-1}. The other end of the spring is held by a frictionless pivot. The mass moves in a circular orbit with a radius of 2 m.
    (a) Calculate the linear velocity of the mass.
    (b) Calculate the angular momentum of the mass about the pivot.

(c) The mass is struck a sudden blow, giving it instantaneous velocity of 1 m/s radially outward. Its tangential velocity is unchanged by the blow. Calculate the total energy of the spring mass system immediately after the blow.

(d) Determine the minimum and maximum values of r (the distance from the pivot) in its new orbit using the Newton–Raphson method.

3.5. How many real solutions does the equation $4x^3-1-\exp(-x^2) = 0$ have? Give two initial values of x which can be used as inputs for the bisection method. Perform two iterations of the bisection method and obtain an approximate solution to the equation.

3.6. A 20 foot ladder is placed in a straight corridor such that the bottom of the ladder is touching the bottom of the left wall and the top of the ladder is leaning on the right wall. Further down the corridor, a 30 foot ladder is placed such that the bottom of this ladder is touching the bottom of the right wall and the top of the ladder is leaning on the left wall. The point where the two ladders 'cross' (as seen by someone further down the corridor) is 10 feet above the ground.

(a) From the information given above, derive an equation which can be used to determine the width of the corridor. Show all the steps used to derive the equation, including the figure which represents the situation described above. Note: do not try to simplify the equation.

(b) Give two initial guesses for the root (the width of the corridor), which can be used to solve the equation in part (a) by the bisection method. (You are not required to obtain the root).

(c) Using the two values which you have given in part (b), determine the approximate width of the corridor (i.e. implement one iteration of the bisection method).

3.7. A spherical water tank has a radius of 3 m. Numerically determine the depth to which the tank should be filled so that it holds water that has a volume of 30 m^3?

3.8. Consider the iterative and Newton–Raphson root-finding methods. Given an arbitrary function, is it guaranteed that at least one of these methods will converge?

3.9. Suppose that you want to use the Newton–Raphson method to find the cube root of a number N. Derive an expression connecting x_i and x_{i+1}, where x_i and x_{i+1} are the approximate values of the cube root at the ith and $(i+1)$th iterations.

3.10. Consider the following set of equations:
$x^2 + y^2 = 9; \, y = \exp(-x)$.

Assume that you have access to a program which finds the solution to a single equation using the fixed-point iteration method. What is your initial guess for x? Do you think the method will converge for this initial guess? Why? How many solutions do you expect for this system of equations?

3.11. Two identical, massive rods are hinged at one end and connected together by a massless ideal spring, as shown in the figure given below. The natural length of the spring is 1 m.

Assume that $g = 10$ m s^{-2} and gravity acts towards the bottom of the page.

If the system is in static equilibrium, what is the angle between the rods (in degrees)?

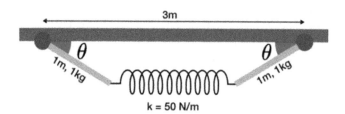

$$3m$$

$$\theta \quad \theta$$

1m, 1kg   1m, 1kg

k = 50 N/m

Set up/derive the relevant equation for the static equilibrium of the system and determine the angle using one of the root-finding methods we discussed in this chapter.

3.12. The potential energy of a particle in a central force scenario is given by $U(r) = -\frac{k}{r} - \frac{C}{r^3} - \frac{D}{r^4}$, where k, C, and D are positive constants. Consider a particle moving with an angular momentum l. For various values of k, C, and D, what is the maximum number of radii at which circular orbits are possible? Hint: Use the concept of the effective potential[9].

3.13. A platform is undergoing simple harmonic motion in a vertical direction with an amplitude of 5 cm and a frequency of $10/\pi$ Hz. A block (whose mass is negligible compared to the mass of the platform) is placed on the platform at the lowest point of its path. Take $g = 1000$ cm s^{-2}.
   (a) At what point will the block leave the platform?
   (b) What is the velocity of the block when it leaves the platform?
   (c) Write down the equations which give the displacements of the block (x_b) and the platform (x_p) as functions of time (after the block leaves the platform). Let $t = 0$ when the block leaves the platform.
   (d) Equating x_b to x_p, determine the approximate time after which the block hits the platform. Use the bisection method and perform two iterations.

3.14. Consider the one-dimensional potential well shown in the figure below. Assume that the total energy of an electron in the well is $E < V_0$.

---

[9] Refer any standard textbook on classical mechanics. For example: *An Introduction to Mechanics* by Kleppner and Kolenkow, 2nd edn, Cambridge University Press, Cambridge, 2013.

(a) Write down the solution to the Schrödinger equation for each of the three regions.

(b) Write down the set of equations that results from applying the boundary conditions.

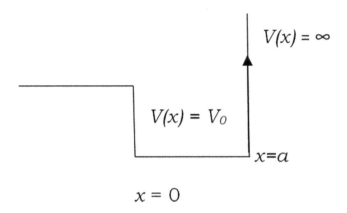

The potential is defined by

$V(x) = V_0$ for $x \leqslant 0$

$V(x) = 0$, for $0 \leqslant x \leqslant a$

$V(x) = \infty$, for $x \geqslant a$.

3.15. How many real solutions do the following equations have? Where is/are the real solutions located?

    (a) $x^7 + 3x^5 + 4x^3 + 5x = 0$

    (b) $\sin(\sin(x)) = 0$

    (c) $\sin(\cos(x)) = 1$

    (d) $\tanh(x^2) = 0$

3.16. How many real solutions does the equation $x^8 + x^6 + x^4 + x^2 + 1 = 0$ have? Explain your reasoning. How many complex solutions will it have?

3.17. Give an example of a function whose root(s) cannot be determined by the bisection method.

    Explain your answer.

3.18. If a polynomial is known to have only complex roots, name the method you will adopt to determine the roots.

3.19. Suppose that you want to use the Newton–Raphson method to find the cube root of a number N. Derive an expression that connects x_i and x_{i+1}, where x_i and x_{i+1} are the approximate values of the cube root at the ith and $(i + 1)$th iterations.

IOP Publishing

Computational Methods Using MATLAB®

An introduction for physicists

P K Thiruvikraman

Chapter 4

Interpolation

On many occasions, we come across tables of values of two variables (say x and y). These are typically determined experimentally[1], and in many cases, the values of x are equally spaced over a certain range. We may now require the value of y corresponding to an x value which is not available in the table (i.e. it is in between two values of x present in the table). In such a situation, we may prefer to determine the value of y by interpolating between the values of y available in the table. This is especially required when the functional dependence of y on x is unknown. For instance, y and x could be the volume and temperature of a non-ideal gas. Since the ideal gas equation of state is not applicable to a non-ideal gas, we are forced to interpolate using the experimentally determined values of V and T.

4.1 Lagrangian interpolation formula

The Taylor series can be used to approximate the value of a function at a point, if the function and its derivatives are known at a nearby point. In practice, the derivatives may not be known and we are therefore forced to explore alternate formulae for interpolation. While many different interpolation formulae are widely known and used, we will confine ourselves to the Lagrangian interpolation formula for the sake of brevity.

The Lagrangian interpolation formula can be obtained from the Taylor series as shown below.

In deriving the Lagrangian interpolation formula, we approximate the function $y = f(x)$ using a polynomial $p(x)$ which matches the function $f(x)$ at a finite number of points. The simplest case is a first-degree polynomial, which will match the function

---

[1] For instance, we may be looking at steam tables which give the volume of steam at different pressures and temperatures.

doi:10.1088/978-0-7503-3791-5ch4

at two points, x_1 and x_2. The first-degree polynomial, i.e. a straight line, is obtained by joining the points $(x_1, f(x_1))$ and $(x_2, f(x_2))$. The slope of this straight line is given by:

$$p'(x) = \frac{p(x_1) - p(x)}{x_1 - x}, \tag{4.1}$$

where we are trying to find the value of $f(x)$ based on the value of the function at the two neighbouring points x_1 and x_2. We can also write the slope of the straight line as:

$$p'(x) = \frac{p(x_2) - p(x)}{x_2 - x}. \tag{4.2}$$

Equations (4.1) and (4.2) can be rearranged and written as:

$$p(x_1) = p(x) + (x_1 - x)p'(x) \tag{4.3}$$

and

$$p(x_2) = p(x) + (x_2 - x)p'(x). \tag{4.4}$$

We can eliminate $p'(x)$ between (4.3) and (4.4) and determine $p(x)$. To eliminate $p'(x)$, multiply (4.3) by (x_2-x) and (4.4) by (x_1-x). Subtract the second of the resulting equations from the first. We obtain:

$$(x_2 - x)p(x_1) - (x_1 - x)p(x_2) = (x_2 - x)p(x) - (x_1 - x)p(x). \tag{4.5}$$

This can be rewritten as:

$$p(x) = \frac{(x_2 - x)p(x_1)}{(x_2 - x_1)} + \frac{(x_1 - x)p(x_2)}{(x_1 - x_2)}. \tag{4.6}$$

Since $p(x_1) = f(x_1)$ and $p(x_2) = f(x_2)$, (4.6) can be rewritten as:

$$p(x) = \frac{(x_2 - x)f(x_1)}{(x_2 - x_1)} + \frac{(x_1 - x)f(x_2)}{(x_1 - x_2)}. \tag{4.7}$$

Equation (4.7) is convenient, as we can approximate the function $f(x)$ by its value at neighbouring points and do not need to know the values of its derivatives. Greater accuracy in interpolation can be achieved by approximating $f(x)$ using a second-degree polynomial. To determine the Lagrangian interpolation of a second-degree polynomial, we need to include the terms up to the second derivative in equation (4.3), which is nothing more than the Taylor series expansion of $p(x)$. We note that $p(x)$ is simply the Taylor series expansion of $f(x)$ truncated up to the term corresponding to the nth derivative.

If we approximate $f(x)$ using a quadratic polynomial, then the two equations (4.3) and (4.4), will be replaced by three equations, each of which has three terms (i.e. up to the second derivative). Instead of going through some tedious algebra to eliminate the first and second derivatives from these three equations, we can use the symmetry in (4.7) to guess the form of the quadratic polynomial. Using the symmetry of (4.7), we arrive at:

$$p(x) = \frac{(x_2 - x)(x_3 - x)f(x_1)}{(x_2 - x_1)(x_3 - x_1)} + \frac{(x_1 - x)(x_3 - x)f(x_2)}{(x_1 - x_2)(x_3 - x_2)} + \frac{(x_1 - x)(x_2 - x)f(x_3)}{(x_1 - x_3)(x_2 - x_3)}. \quad (4.8)$$

Using this symmetry argument, we can obtain the Lagrangian interpolation of a polynomial without recourse to any algebra!

Based on (4.8), we can write the general form for the interpolation of a polynomial of degree n, as follows:

$$p_n(x) = \sum_{j=1}^{N} l_{j,N} f(x_j) \quad (4.9)$$

where

$$l_{j,N}(x) = \frac{(x - x_1)(x - x_2).....(x - x_N)}{(x_j - x_1)(x_j - x_2)....(x_j - x_3)} = \frac{\prod_{i=1}^{N}(x - x_i)}{\prod_{i=1}^{N}(x_j - x_i)} \quad \text{where } i \neq j. \quad (4.10)$$

Example 4.1

Given that $\ln(1) = 0$, $\ln(2) = 0.69315$, and $\ln(5) = 1.60944$, use Lagrangian polynomial interpolation to determine $\ln(3)$.

Solution

Since three points have been given, we can use a second-degree Lagragian polynomial which passes through these three points; (4.8) gives the quadratic Lagrangian polynomial, which is what we require here.

$$\ln(3) = \frac{(3 - 2)(3 - 5)}{(1 - 2)(1 - 5)} \times 0 + \frac{(3 - 1)(3 - 5)}{(2 - 1)(2 - 5)} \times 0.69315$$

$$+ \frac{(3 - 1)(3 - 2)}{(5 - 1)(5 - 2)} \times 1.60944 = 1.0986.$$

This is the exact value of $\ln(3)$, accurate to four decimal places!

Suppose we had only used the values at $x = 2$ and $x = 5$ to obtain the value at $x = 3$; in this case, we would have obtained:

$$\ln(3) = \frac{(3 - 5)}{(2 - 5)} \times 0.69315 + \frac{(3 - 2)}{(5 - 2)} \times 1.60944 = 0.9986.$$

We can see that using a quadratic polynomial leads to better results than those obtained by linear interpolation. Should we use a very high-degree polynomial to obtain more accuracy? In most cases, the accuracy of the interpolation increases when we increase the degree of the polynomial. However, note that increasing the order of the polynomial leads to an increase in the number of computations. Further, the oscillations in a polynomial of a very high degree can increase the error in some cases.

4.2 The error caused by interpolation

Before we move on to discuss another technique for interpolation, let us try to estimate the error involved in approximating a function $f(x)$ using a polynomial $p_n(x)$ of degree n.

We start by noting that the error itself will be a function of x, as the polynomial of degree n will match the value of the function at $n+1$ points. Therefore the error function will be equal to zero at those $n+1$ points, while it may be nonzero elsewhere.

The error function is, by definition, the difference between $f(x)$ and $p_n(x)$. Furthermore, using the fact that the polynomial matches the function at $n+1$ points, the error function $E(x)$ is equal to:

$$E(x) = f(x) - p_n(x) = (x - x_0)(x - x_1)(x - x_2)(x - x_3)\ldots\ldots(x - x_n)g(x), \quad (4.11)$$

where x_0, x_1, ..., x_n are the $n+1$ points at which the polynomial is equal to the function; $g(x)$ determines at the other points where the error will, in general, be nonzero.

Assuming that $E(x)$ is a continuous function, we can see that, since the error is zero at $n+1$ points, its derivative will be zero at n points. Consider the auxiliary equation:

$$w(t) = f(t) - p_n(t) - (t - x_o)(t - x_1)(t - x_2)\ldots..(t - x_n)g(x); \quad (4.12)$$

$w(t)$ is zero at $t = x_0, x_1, x_2,\ldots, x_n$ and also at $t = x$.

Therefore, $w(t)$ has $n+2$ zeros. Between two zeros of $w(t)$ there must be one zero of $w'(t)$. Since $w(t)$ has $n+2$ zeros, $w'(t)$ has $n+1$ zeros, and $w'(t)$ has n zeros. Proceeding along similar lines, we can see that the $(n+1)$th derivative has one zero, which will be at $t = \xi$ (all we know about this point is that it is some point which is in the interval $x_0 < \xi < x_n$).

Therefore, we have:

$$w^{(n+1)}(\xi) = 0 = \frac{d^{n+1}}{dt^{n+1}}\big[f(t) - p_n(t) - (t - x_o)(t - x_1)(t - x_2)\ldots..(t - x_n)g(x)\big]. \quad (4.13)$$

Since $p_n(t)$ is a polynomial of the nth degree, its $(n+1)$th derivative will be zero, and the last term in (4.13) involves a polynomial of degree $n+1$ in t, which yields $(n+1)!$ on being differentiated with respect to t. Here, $g(x)$ will be treated as a constant, since it is only a function of x and not t. Therefore, (4.13) can be rewritten as:

$$w^{(n+1)}(\xi) = 0 = f^{(n+1)}(\xi) - (n + 1)!g(x). \quad (4.14)$$

Defining the auxiliary function has helped us to determine $g(x)$, and therefore the error function, from (4.14):

$$g(x) = \frac{1}{(n + 1)!} \frac{d^{n+1}f(t)}{dt^{n+1}}\bigg|_{t=\xi}. \quad (4.15)$$

However, (4.15) is of limited use in practice, as we normally do not know the form of $f(t)$ when interpolating within a table of values, and the point $t = \xi$ is also unknown.

Still, (4.15) does have its uses, as we will see when we discuss techniques for numerical integration.

4.3 Newton's form of interpolation polynomial

An alternative formulation for interpolation is provided by Newton's interpolation polynomial, which uses 'divided' differences. This polynomial has the form:

$$p_n(x) = a_0 + a_1(x - x_0) + a_2(x - x_0)(x - x_1)$$

$$+ \ldots\ldots a_n(x - x_0)(x - x_1)\ldots\ldots(x - x_n) = \sum_{i=0}^{n} a_i \left(\prod_{j=0}^{i-1} (x - x_j) \right). \tag{4.16}$$

The particular structure of Newton's interpolation polynomial means that (4.16) leads to a system of equations of the form:

$$a_0 = f(x_0)$$
$$a_0 + a_1(x_1 - x_0) = f(x_1) \tag{4.17}$$
$$a_0 + a_1(x_2 - x_0) + a_2(x_2 - x_0)(x_2 - x_1) = f(x_2)$$

and so on.

The solution to the $n+1$ equations listed above can be obtained by forward substitution:

$$a_1 = \frac{f(x_1) - f(x_0)}{x_1 - x_0} = \frac{f[x_1, x_0]}{x_1 - x_0} \tag{4.18}$$

$$a_2 = \frac{f[x_2, x_1] - f[x_1, x_0]}{x_2 - x_0}.$$

With $n+1$ interpolation points, n columns need to be computed to form the complete divided difference table. For $i = 1, 2, 3,\ldots, n$, the ith column contains $n+1-i$ values that need to be calculated, each of which requires two subtractions and one division. Therefore, the number of operations required to compute the entire table is:

$$3 \sum_{i=1}^{n} (n + 1 - i) = 3n^2 + 3n - \frac{3n(n + 1)}{2} = \frac{3n(n + 1)}{2}. \tag{4.19}$$

Note that (4.19) only gives us the number of computations required to arrive at the constants $a_0, a_1, a_2, \ldots, a_n$ in (4.16). We would still need to perform subtractions, additions, and multiplications to evaluate all the terms in (4.16) and use it to arrive at the value of the function at x.

The number of multiplications involved in (4.16) will be $1 + 2 + 3 + \ldots, n = n(n + 1)/2$

The number of additions (including subtractions)

$$= n + 1 + 2 + 3 + \ldots.n = n(n + 3)/2 \tag{4.20}$$

These have to be added to the number of operations required to compute the table (4.19) to arrive at the total number of computations. However, note that each time we try to interpolate at a new point x, we only need to perform the calculations in (4.20). The calculations required to compute the table only have to be performed once.

Example 4.2

Determine Newton's form of the interpolation polynomial for the following dataset, then use this polynomial to estimate y when $x = 1.5$.

x	−1	0	1	2
y	−5	1	2	10

x	Y	$\Delta y/\Delta x$	$\Delta^2 y$	$\Delta^3 y$
−1	−5			
0	1	6/1 = 6		
1	2	2/1 = 2	−4/2 = −2	
2	10	8/1 = 8	6/2 =3	5/3

Using the table and (4.16), we have:

$$f(1.5) = f(-1) + 6*(1.5 - (-1)) + (-2)*(1.5 - (-1))(1.5 - 0)$$
$$+ 5/3(1.5 - (-1))(1.5 - 0)(1.5 - 1) = -5 + 15 - 7.5 + 3.125 = 5.625.$$

Having completed this numerical example, let us now compare Newton's interpolation polynomial to that of Lagrange. The main advantage of Netwon's polynomial is that the table is computed only once, while all the calculations have to be repeated in the case of the Lagrangian interpolation polynomial whenever we interpolate at a new point.

We note that each of the $n + 1$ terms in (4.11) has n multiplications in the numerator and $n-1$ multiplications in the denominator. We also need to divide the numerator by the denominator. Therefore the total number of multiplications is $(2n)(n + 1)$. Within each of the $n + 1$ terms to be summed in (4.11), we have n subtractions in the numerator and n subtractions in the denominator. Furthermore, we need to sum $n + 1$ separate terms. The total number of additions is $(n + 1)(2n) + n = 2n^2 + 3n$.

Clearly, the number of calculations required for the Lagrangian interpolation polynomial is greater than the number required for Newton's polynomial (4.20), though both are of the order of n^2.

Exercises

4.1 If $L_n(x)$ is the Lagrangian interpolation of degree n and $P_n(x)$ is Newton's interpolation polynomial (also of degree n) for a function $f(x)$, which one of them will have a smaller error (or will they have same error), when they are used to interpolate and determine the value of the function at a point x. The same set of $n+1$ points $(x_0, x_1, ..., x_n)$ was used to determine both of the interpolation polynomials.

4.2 Suppose that you are trying to find the value of $\exp(x)$ by interpolation using the values of this function at three points: x_o, x_1, and x_2. You can use the Lagrangian interpolation polynomial . Alternatively, you can use the series expansion for $\exp(x)$, $\exp(x) = 1 + x + x^2/2 + ...$ and truncate it to a quadratic polynomial. Does the quadratic Lagrangian interpolation polynomial have any particular advantage(s), compared to the series expansion mentioned above?

4.3 Write a program to calculate the value of a function at a particular x using the Lagrangian interpolation polynomial, given the value of the function at $n + 1$ points.

4.4 Write a program to construct the divided difference table and hence interpolate and arrive at the value of a function at a particular x using Newton's form of the polynomial.

4.5 Given that $\ln(9) = 2.1972$ and $\ln(9.5) = 2.2513$, obtain the value of $\ln(9.3)$ using a Lagrangian interpolation polynomial . Obtain a numerical estimate for the error that arises when this Lagrangian interpolation polynomial is used.

4.6 Given the four points (2,1), (4), (3,5), and (8,9), find the cubic Lagrangian polynomial that passes through them. Use this polynomial to find the value of y for $x = 5$.

4.7 Suppose you are given a table of values for a function $f(x)$ versus x. Which of the following procedures is likely to yield a more accurate numerical value for the first derivative of the function at one of the tabulated values, say x_1?

   (a) Use the forward difference formula to compute the first derivative at x_1.

   (b) Fit the given data to a polynomial $p(x)$ (degree >2), then differentiate $p(x)$ and obtain the derivative at x_1.

IOP Publishing

Computational Methods Using MATLAB®

An introduction for physicists
P K Thiruvikraman

Chapter 5

Numerical linear algebra

The tools of linear algebra are used widely in physics. On numerous occasions (e.g. in Newtonian mechanics, circuit analysis, etc.), we need to solve a system of linear equations. If the number of equations is, say, two or three, we can easily solve them by hand using elimination or substitution. However, if we have many equations (for example, when analysing a complicated circuit), a systematic programmable procedure that solves this equation will be helpful.

We also come across situations wherein we need to determine the eigenvalues, eigenvectors, or the determinant of a matrix. The usual pen-and-paper methods used to determine eigenvalues and eigenvectors are not designed for automation; hence, new techniques have been invented for this purpose.

In this chapter, we will look at the burgeoning field of numerical linear algebra, which should form part of the armoury of a computational physicist.

5.1 Solving a system of equations: Gaussian elimination

Consider the system of n equations given below:

$$a_{11}x_1 + a_{12}x_2 + \ldots a_{1N}x_N = b_1$$

$$a_{21}x_1 + a_{22}x_2 + \ldots a_{2N}x_N = b_2$$

$$\ldots\ldots\ldots\ldots\ldots\ldots\ldots\ldots\ldots\ldots\ldots\ldots\ldots\ldots\ldots\ldots\ldots\ldots$$
$$\ldots\ldots\ldots\ldots\ldots\ldots\ldots\ldots\ldots\ldots\ldots\ldots\ldots\ldots\ldots\ldots\ldots\ldots$$

$$a_{N1}x_1 + a_{N2}x_2 + \ldots a_{NN}x_N = b_N.$$

How do we systematically solve this system of equations? If N is very large, then we definitely need a systematic programmable procedure, as the substitution or elimination of some variables chosen at random will not allow us to keep track of our operations. When we are trying to eliminate some variable, we take linear combinations of two or more equations from this set. In subsequent operations, we have to ensure that we are always dealing with a set of linearly independent equations.

The technique of Gaussian elimination helps us to systematise the procedure of elimination and substitution. In Gaussian elimination, we eliminate the variable x_1 from all equations except the first, we eliminate the variable x_2 from all equations except the first two, and so on. We now illustrate Gaussian elimination by a numerical example.

Example 5.1

Solve the following system of equations:

$$2x + y + z = 5$$
$$4x - 6y = -2 \tag{5.1}$$
$$2x + 7y + 2z = 9.$$

To apply Gaussian elimination, we form the augmented matrix made up of the coefficients of the variables in the above equations and the right-hand sides of these equations:

$$\begin{bmatrix} 2 & 1 & 1 & 5 \\ 4 & -6 & 0 & -2 \\ -2 & 7 & 2 & 9 \end{bmatrix}. \tag{5.2}$$

Note that the last column of this matrix is formed using the constant terms on the right-hand sides of these equations. Each row of the matrix is obtained from the corresponding equation in (5.1). To apply the systematic elimination procedure described earlier, we divide the first equation (or the first row of (5.2)) by two (this is known as the pivot) and multiply it by four. The result is subtracted from the second row and replaces the second row. In other words, we perform the row operation: $R_2 \rightarrow R_2 - \frac{4}{2}R_1$. Similarly, $R_3 \rightarrow R_3 - \frac{-2}{2}R_1$. Performing these two row operations leads us to the following matrix:

$$\begin{bmatrix} 2 & 1 & 1 & 5 \\ 0 & -8 & -2 & -12 \\ 0 & 8 & 3 & 14 \end{bmatrix}. \tag{5.3}$$

We now proceed to similarly make the second term in the last row of (5.3) zero. Performing the row operation $R_3 \rightarrow R_3 - \frac{-8}{8}R_2$, we get:

$$\begin{bmatrix} 2 & 1 & 1 & 5 \\ 0 & -8 & -2 & -12 \\ 0 & 0 & 1 & 2 \end{bmatrix}. \tag{5.4}$$

Since each row of (5.4) corresponds to an equation, the last row of (5.4) should stand for the equation $z = 2$. We now substitute this into the second equation to obtain the value of y:

$$y = \frac{-12 + 2z}{-8} = 1. \tag{5.5}$$

Similarly, we can substitute the values of z and y into the first equation to obtain the value of x:

$$x = \frac{5 - z - y}{2} = 1. \tag{5.6}$$

To automate the above procedure, we need to write down general expressions for these row operations. We note from equations (5.3) to (5.6) that the pivot is an element that is on the diagonal of the augmented matrix, and in the kth iteration, the pivot is A_{kk}, where A is the augmented matrix. At the kth iteration, an element $A(i,j)$ of the augmented matrix is modified in the following manner:

$$A(i, j) = A(i, j) - \frac{A(i, k)}{A(k, k)} A(k, j) \tag{5.7}$$

A MATLAB program that implements elimination (5.7) and back-substitution is shown below:

```
'Gaussian elimination';
'simultaneous equations to be entered as a matrix';
'RHS of equations to be entered in the last column';
clear;
a=[2 1 1 5;4 -6 0 -2;-2 7 2 9];
n=size(a,1);
'n is the number of equations, equal to number of rows of a';
'elimination';
k=2;
b=a;
while k<n+1
for i=k:n
    for j=k-1:n+1

        b(i,j)=a(i,j)-a(k-1,j)*a(i,k-1)/a(k-1,k-1);
        'This is the implementation of (5.7), but here k-1 is used as k
was initially set to 2, so the iteration number is actually k-1';

    end
end
k=k+1;
a=b;
end
x(n)=a(n,n+1)/a(n,n);
'back-substitution';
k=n;
for i=n-1:-1:1
    sum=0;
    for j=n:-1:k
        sum=sum+a(i,j)*x(j);
    end
    x(i)=(a(i,n+1)-sum)/a(i,k-1);
    k=k-1;
end
```

In the program above, the vector x contains the solution to the equations.

The above program also helps us to determine the number of additions and multiplications involved in Gaussian elimination. By convention, subtraction is treated as an addition and division is treated as a multiplication. If there are n equations, we need to implement equation (5.7) $n-1$ times during the first iteration. It is more efficient to first calculate A $(i,k)/A(k,k)$ in (5.7), since this is a constant for a single equation, and then multiply this by each term in the ith equation.

During the first iteration, we need to calculate the ratio $A(i,k)/A(k,k)$, $(n-1)$ times; thus, the number of multiplications is $(n+1)(n-1)$. Therefore the number of multiplications is $(n-1)(n+2)$. In the next iteration, the number of multiplications will be $(n-2)(n+1)$, since the number of terms in the equation has been reduced by one (the first term having been reduced to zero). Therefore, in each successive iteration, the number of multiplications can be obtained by replacing n by $n-1$ in the number of multiplications used in the previous iteration. Therefore, the total number of multiplications for all the iterations combined is:

$$(n-1)(n+2) + (n-2)(n+1) + \ldots.1\times4 = \sum_{k=1}^{n-1} k(k+3). \tag{5.8}$$

The summation in (5.8) can be easily evaluated using the standard results for the sum of the first $(n-1)$ integers and the sum of the squares of the first $(n-1)$ integers, to give:

$$\sum_{k=1}^{n-1} k(k+3) = \sum_{k=1}^{n-1} k^2 + \sum_{k=1}^{n-1} 3k = \frac{(n-1)(n)(2n-1)}{6} + \frac{3n(n-1)}{2} \approx O(n^3). \tag{5.9}$$

Apart from the multiplications, we need to perform subtractions in (5.7). We notice that for the first iteration of (5.7), we need to perform $(n+1)(n-1)$ additions, since there are $(n-1)$ equations and each equation has $(n+1)$ terms. Therefore, the total number of subtractions is:

$$(n-1)(n+1) + (n-2)(n) + \ldots.1\times3 = \sum_{k=1}^{n-1} k(k+2). \tag{5.10}$$

Equation (5.10) is very similar to (5.8), so the summation in (5.10) can be evaluated similarly:

$$\sum_{k=1}^{n-1} k(k+2) = \sum_{k=1}^{n-1} k^2 + \sum_{k=1}^{n-1} 2k = \frac{(n-1)(n)(2n-1)}{6} + n(n-1) \approx O(n^3). \tag{5.11}$$

The algorithm mentioned above for Gaussian elimination has some pitfalls. In particular, what happens if one of the pivots becomes zero? In the next iteration, the program would try to divide the elements of a row by zero, leading to an error. To avoid this, we need to swap two rows, as shown in the following example.

Example 5.2

Solve the following system of equations by Gaussian elimination:

$$\begin{aligned} x + y + z &= -2 \\ 3x + 3y - z &= 6 \\ x - y + z &= -1. \end{aligned} \tag{5.12}$$

Following the procedure of example 5.1, we start by writing down the augmented matrix. The augmented matrix for the above set of equations is:

$$\begin{bmatrix} 1 & 1 & 1 & -2 \\ 3 & 3 & -1 & 6 \\ 1 & -1 & 1 & -1 \end{bmatrix}. \tag{5.13}$$

We perform row operations to replace the second and third rows. The augmented matrix now becomes:

$$\begin{bmatrix} 1 & 1 & 1 & -2 \\ 0 & 0 & -4 & 12 \\ 0 & -2 & 0 & 1 \end{bmatrix}. \tag{5.14}$$

We note that the pivot is equal to zero in the second row. We also note that dividing by zero can be avoided by swapping the second and third rows (which is equivalent to swapping the second and third equations in (5.13)).

Swapping the second and third rows in (5.14), we get:

$$\begin{bmatrix} 1 & 1 & 1 & -2 \\ 0 & -2 & 0 & 1 \\ 0 & 0 & -4 & 12 \end{bmatrix}. \tag{5.15}$$

We can now determine the solution by back-substitution to obtain $z = -3$, $y = -1/2$, and $x = 3/2$.

The reader is encouraged to modify the MATLAB program provided earlier to handle the case in which the pivot becomes zero. In some cases, the swapping of adjacent rows may not be enough, and we might have to swap non-adjacent rows to prevent the pivot from becoming zero. If the original system of equations is inconsistent, there is no solution, and swapping will not solve the problem. If the last equation in (5.12) had been $x+y+z = -1$, then, clearly, the third equation would have been inconsistent with the first equation, which would mean that no consistent solution exists for this system of equations.

The MATLAB function rref(A) reduces the matrix A to row echelon form.

5.2 Evaluating the determinant of a matrix

It is assumed that the reader has completed an elementary course in linear algebra[1], in which he has encountered the determinant of a matrix. In introductory courses, students are taught to use the Laplace expansion to evaluate the determinant of a matrix. The determinant usually has to be evaluated in order to solve a system of equations using Kramer's rule or to determine the inverse of a matrix. We will quickly review the procedure used to evaluate the determinant[2] using the Laplace expansion.

Example 5.3

Evaluate the determinant of the coefficient matrix in (5.1) using the Laplace expansion. The matrix formed by the coefficient in (5.1) is:

$$\begin{bmatrix} 2 & 1 & 1 \\ 4 & -6 & 0 \\ -2 & 7 & 2 \end{bmatrix}. \tag{5.16}$$

The determinant of this matrix is represented as:

$$\begin{vmatrix} 2 & 1 & 1 \\ 4 & -6 & 0 \\ -2 & 7 & 2 \end{vmatrix}. \tag{5.17}$$

The Laplace expansion of this determinant is:

$$2\begin{vmatrix} -6 & 0 \\ 7 & 2 \end{vmatrix} - \begin{vmatrix} 4 & 0 \\ -2 & 2 \end{vmatrix} + \begin{vmatrix} 4 & -6 \\ -2 & 7 \end{vmatrix} = -24 - 8 + (28 - 12) = -16. \tag{5.18}$$

While you may think that I wasted your time in example 5.3 by illustrating a procedure which you already knew, I deliberately ran through all the steps so that we can easily enumerate the number of additions and subtractions that have to be performed for the Laplace expansion of a determinant. We note from (5.17) and (5.18) that in the Laplace expansion, we write a determinant of order n as the sum of n determinants, each of order $n-1$.

Let us denote the number of multiplications required to evaluate a determinant of order n as $m(n)$, and let $a(n)$ denote the number of additions needed to evaluate the same determinant using the Laplace expansion. From (5.18), we can generalize to the case of a determinant of order and say that:

$$m(n) = n \times m(n - 1) + n. \tag{5.19}$$

In (5.18) we multiplied the three cofactors by two, minus one, and one because we were dealing with a determinant of order three. This generalises to n multiplications

---

[1] See, for instance, *Linear Algebra and Its Applications* by Gilbert Strang, 4th edn, Thomson, Belmont, CA, 2007, which provides an excellent introduction to Linear algebra.
[2] The MATLAB function det(A) yields the determinant of the square matrix A.

for a determinant of order n. In addition to this, we need to perform multiplications to evaluate each of the n determinants of order $n-1$.

In turn, each of these determinants of order $n-1$ will require:

$$m(n-1) = (n-1) \times m(n-2) + n - 1 \tag{5.20}$$

multiplications.

Substituting (5.20) into (5.19), we obtain:

$$m(n) = n[(n-1)m(n-2) + (n-1)] + n. \tag{5.21}$$

By continuing this recursive process until we reach a determinant of order two, we obtain:

$$m(n) = n[(n-1)[(n-2)m(n-3) + n - 2] + (n-1)] + n \tag{5.22}$$

$$m(n) = n(n-1)(n-2).....3.2 + n(n-1)(n-2).....3 + n(n-1)(n-2)....4 + ... \tag{5.23}$$

We can rewrite (5.23) as:

$$m(n) = n! + \frac{n!}{2!} + \frac{n!}{3!} + ... = n!\left(1 + \frac{1}{2!} + \frac{1}{3!} + ...\right) = n!(e - 1). \tag{5.24}$$

Adopting the same recursive approach to enumerate the number of additions, we see that:

$$a(n) = n \times a(n-1) + n - 1 \tag{5.25}$$

$$a(n) = n[(n-1)a(n-2) + (n-2)] + n - 1 \tag{5.26}$$

$$a(n) = n! + n - 1 + n(n-2) + n(n-1)(n-3) + . \tag{5.27}$$

We can see from (5.24) and (5.27) that the numbers of additions and multiplications are both of the order of $n!$. The factorial of n increases very rapidly with increasing n. In fact, from Stirling's formula[3] (which is valid for large n), we obtain:

$$n! = n^n \exp(-n)\sqrt{2\pi n}. \tag{5.28}$$

To get a feel for the rapid increase of $n!$, look at the numbers given in table 5.1. The table also shows that the error in Stirling's formula is smaller for large values of n.

Equations (5.24) and (5.27) show that the Laplace expansion is computationally costly for large n. Therefore, we now discuss a technique known as lower–upper (LU) decomposition, which makes it easy to evaluate the determinant of a matrix and also to evaluate its inverse.

---

[3] Stirling's formula is discussed in most books on mathematical physics. See for instance, *Mathematical Methods in the Physical Sciences*, by Mary L. Boas, 3rd edn, Wiley, New York, 2006.

Table 5.1 n, $n!$, and $n!$ from Stirling's formula together with the error in Stirling's formula.

N	$n!$	Stirling's formula	Error %
1	1	0.922137	7.786299
2	2	1.919004	4.049782
3	6	5.83621	2.72984
4	24	23.50618	2.057604
5	120	118.0192	1.650693
6	720	710.0782	1.37803
7	5040	4980.396	1.182622
8	40320	39902.4	1.035726
9	362880	359536.9	0.921276
10	3628800	3598696	0.829596
11	39916800	39615625	0.754507
12	4.79E+08	4.76E+08	0.691879
13	6.23E+09	6.19E+09	0.63885
14	8.72E+10	8.67E+10	0.59337
15	1.31E+12	1.3E+12	0.553933
16	2.09E+13	2.08E+13	0.519412
17	3.56E+14	3.54E+14	0.48894
18	6.4E+15	6.37E+15	0.461846
19	1.22E+17	1.21E+17	0.437596
20	2.43E+18	2.42E+18	0.415765

5.3 LU decomposition

We now discuss a technique known as LU decomposition, which facilitates the evaluation of the determinant and the inverse of a matrix.

The LU decomposition of a matrix A can be written as:

$A = LU$

Here, L stands for a lower triangular matrix, while U is an upper triangular matrix. Furthermore, all the diagonal elements[4] of U are one.

Before we proceed further, let us demonstrate the advantages of this decomposition. We note that the determinant of A should be equal to the product of the determinants of L and U. Furthermore, the determinant of U should be equal to one (prove this using the Laplace expansion of U). From the Laplace expansion of L, we see that its determinant is equal to the product of its diagonal elements. Therefore, once we complete the LU decomposition, we just require n multiplications (and no additions) to compute the determinant of L (hence that of A). Of course, we need to

---

[4] It is also possible to have a decomposition in which, instead of U, the diagonal elements of L are unity. Other decompositions, such as the QR decomposition, are also possible.

do some computation to achieve the LU decomposition, but, as we will show, this is much smaller than the number of operations required for the Laplace expansion.

We now arrive at the equations for the LU decomposition by noting that the equation $A = LU$ implies:

$$\begin{pmatrix} a_{11} & a_{12} & a_{13} & ..... & a_{1n} \\ a_{21} & a_{22} & a_{23} & ...... & a_{2n} \\ ...... & ...... & ...... & ...... & ...... \\ ...... & ...... & ...... & ...... & ...... \\ a_{n1} & a_{12} & a_{n3} & ...... & a_{nn} \end{pmatrix} = \begin{pmatrix} l_{11} & 0 & 0 & ..... & 0 \\ l_{21} & l_{22} & 0 & ...... & 0 \\ ...... & ...... & ...... & ...... & ...... \\ ...... & ...... & ...... & ...... & ...... \\ l_{n1} & l_{12} & l_{n3} & ...... & l_{nn} \end{pmatrix}$$

$$\begin{pmatrix} 1 & u_{12} & u_{13} & ..... & u_{1n} \\ 0 & 1 & u_{23} & ...... & u_{2n} \\ ...... & ...... & ...... & ...... & ...... \\ ...... & ...... & ...... & ...... & ...... \\ 0 & 0 & 0 & ...... & 1 \end{pmatrix}.$$

By carrying out the multiplication of the matrices L and U and equating each element of the product matrix to the corresponding element of A, we get:

$$a_{11} = l_{11} \tag{5.29}$$

$$a_{12} = l_{11} u_{12} \tag{5.30}$$

or, in general,

$$a_{1j} = l_{11} u_{1j}. \tag{5.31}$$

From (5.31) and (5.29) above, we can determine the elements in the first row of the matrix U:

$$u_{1j} = \frac{a_{1j}}{l_{11}} = \frac{a_{1j}}{a_{11}}. \tag{5.32}$$

Similarly, we see that:

$$a_{21} = l_{21}. \tag{5.33}$$

Generalizing from (5.29) and (5.33):

$$a_{i1} = l_{i1}. \tag{5.34}$$

Proceeding to the other elements, we see that:

$$a_{22} = l_{21} u_{12} + l_{22}. \tag{5.35}$$

Since u_{12} is already determined by (5.32) and l_{21} is determined by (5.33), we can now use (5.35) to determine l_{22}; thus, the general principle is that we can evaluate the elements of the L and U matrices using the elements of these very matrices which appear in an earlier row or column. From (5.35), we get:

$$l_{22} = a_{22} - l_{21}u_{12}. \tag{5.36}$$

This can be generalised to:

$$l_{ik} = a_{ik} - \sum_{j=1}^{k-1} l_{ij}u_{jk}. \tag{5.37}$$

A general expression for the elements of the U matrix can be obtained by noting that:

$$a_{23} = l_{21}u_{13} + l_{22}u_{23}. \tag{5.38}$$

Equation (5.38) can be rewritten as:

$$u_{23} = \frac{a_{23} - l_{21}u_{13}}{l_{22}}. \tag{5.39}$$

The generalization of (5.39) is:

$$u_{kj} = \frac{a_{kj} - \sum_{i=1}^{k-1} l_{ki}u_{ij}}{l_{kk}}. \tag{5.40}$$

Note that (5.37) and (5.40) have to be used interchangeably, as the elements of L and U matrix depend on each other.

The program given below shows the implementation of the LU decomposition:

```
'LU decomposition of a matrix';
clear;

a=[2 1 1;4 -6 0;-2 7 2];
'a is the matrix which we wish to decompose in the form LU';
rows=size(a,1);
cols=size(a,2);
'generating the first column of l, which will be same as the
first column of the original matrix';
for i=1:rows
    l(i,1)=a(i,1);
end
```

```
'setting elements above the diagonal to zero in the lower
triangular matrix';
for i=1:rows
    for j=i+1:cols
        l(i,j)=0;
    end
end
'Setting the diagonal elements of matix U equal to 1';
for i=1:rows
    u(i,i)=1;
end
'Setting the elements below the diagonal equal to zero in the
upper triangular matric';
for i=2:rows
    for j=1:i-1
        u(i,j)=0;
    end
end
'generating the first row of matrix U';
for i=2:cols
    u(1,i)=a(1,i)/l(1,1);
end
'generating the remaining elements of matrices U and L';
for i=2:rows
    for j=2:i
        sum=0;
        for k=1:i-1
        sum=sum+l(i,k)*u(k,j);
        end
        l(i,j)=a(i,j)-sum;
    end
    for j=i+1:cols
        sum=0;
        for k=1:i-1
            sum=sum+l(i,k)*u(k,j);
        end
        u(i,j)=(a(i,j)-sum)/l(i,i);
    end
end
```

The input matrix for the above program (given at the beginning of the program) was:

$$\begin{pmatrix} 2 & 1 & 1 \\ 4 & -6 & 0 \\ -2 & 7 & 2 \end{pmatrix}.$$

If you run the above program, you will find that the L and U matrices are:

$$L = \begin{pmatrix} 2 & 0 & 0 \\ 4 & -8 & 0 \\ -2 & 8 & 1 \end{pmatrix} \quad U = \begin{pmatrix} 1 & 0.5 & 0.5 \\ 0 & 1 & 0.25 \\ 0 & 0 & 1 \end{pmatrix}.$$

The three nested 'for' loops in the above program show that, ignoring the initialisation of the arrays, the number of operations involved in the LU decomposition is of the order of N^3 for an $N \times N$ matrix.

5.4 Determination of eigenvalues and eigenvectors: the power method

Many physical situations demand the determination of eigenvalues and eigenvectors of a matrix. For instance, if we have a system of coupled oscillators, we may wish to find the normal-mode frequencies of the system. Any mode of vibration of such a system can be written as a linear combination of the vibrations of the normal modes. Representing the motion of the system in terms of the normal modes is very useful, as the normal modes are independent from each other, and the energy in one normal mode cannot be exchanged with that of another normal mode (even though different parts of the system can exchange energy, as they are coupled).

The eigenvalue equation for a matrix A is represented by $Ax = \lambda x$. The word 'eigen' means 'characteristic' in the German language. Here, λ is known as an eigenvalue and the corresponding eigenvector is x. The above eigenvalue equation can be rewritten as $(A - \lambda I)x = 0$. For the eigenvector x to be nonzero, we must have: $|A - \lambda I| = 0$. The condition mentioned above, in which the determinant vanishes, helps us to determine the eigenvalues and hence the eigenvectors. We can see that expanding the above determinant leads to a polynomial in λ. The roots of this polynomial are the eigenvalues of A. We illustrate this process with a numerical example.

Example 5.4

Determine the eigenvalues and eigenvectors of $A = \begin{pmatrix} 2 & -1 \\ -1 & 1 \end{pmatrix}$. The eigenvalue equation for this matrix leads to the condition whereby $\begin{vmatrix} 2 - \lambda & -1 \\ -1 & 1 - \lambda \end{vmatrix} = 0$. The expansion of this determinant leads to the following quadratic equation:

$$(2 - \lambda)(1 - \lambda) - 1 = 0$$

$$\lambda^2 - 3\lambda + 1 = 0.$$

The two roots of this quadratic equation are: $\lambda = \frac{3 \pm \sqrt{5}}{2}$.

By substituting each eigenvalue into the eigenvalue equation, we can determine the corresponding eigenvectors. For example, substituting the larger eigenvalue into the eigenvalue equation leads to:

$$Ax = \begin{pmatrix} 2 & -1 \\ -1 & 1 \end{pmatrix}\begin{pmatrix} x_1 \\ x_2 \end{pmatrix} = \lambda \begin{pmatrix} x_1 \\ x_2 \end{pmatrix}.$$

This leads to the equation $2x_1 - x_2 = \frac{3 + \sqrt{5}}{2}x_1$ or $\frac{1 - \sqrt{5}}{2}x_1 = x_2$. The other eigenvector can be determined by substituting the corresponding eigenvalue.

Note that while the eigenvalues are unique for a given matrix, the corresponding eigenvectors are not unique. We can only determine the ratio of the components of the eigenvectors. Any vector whose components have the required ratio can be chosen as an eigenvector.

While the numerical calculations in example 5.4 seem straightforward, the method adopted does not easily lend itself to automation. Note that the determinant has to be expanded by hand to arrive at the polynomial in λ. One can use root-finding methods to get the roots, but then expanding the determinant is tedious, especially if the order of the determinant is very large.

The power method, by contrast, can easily be programmed. However, it only determines the largest eigenvalue. This technique can be modified to determine all the remaining eigenvalues. We will first describe the method used to determine the largest eigenvalue. The first step in the power method is to assume or guess the eigenvector which corresponds to the largest eigenvalue. Let us illustrate the power method by using it to determine the largest eigenvalue and eigenvector of matrix A given in example 5.4.

Let the initial guess for the eigenvector be: $x = \begin{pmatrix} 1 \\ 1 \end{pmatrix}$. We now act on this vector using the matrix A. Since the initial guess may not correspond to one of the eigenvectors, the result will be a new vector, x'. We now act on x' using the matrix A and repeat this process. After many iterations, it is found that we converge on the eigenvector corresponding to the largest eigenvalue. We will give a detailed proof of the convergence of the power method, but at the moment, we only present the results of the successive iterations.

$$Ax = \begin{pmatrix} 2 & -1 \\ -1 & 1 \end{pmatrix}\begin{pmatrix} 1 \\ 1 \end{pmatrix} = \begin{pmatrix} 1 \\ 0 \end{pmatrix} = x'$$

$$Ax' = \begin{pmatrix} 2 & -1 \\ -1 & 1 \end{pmatrix}\begin{pmatrix} 1 \\ 0 \end{pmatrix} = \begin{pmatrix} 2 \\ -1 \end{pmatrix} = x'' = 2\begin{pmatrix} 1 \\ -\frac{1}{2} \end{pmatrix}.$$

Before we act on x' using A, we 'pull' out the largest component of the vector x' (this step was not necessary in the previous iteration, as the largest component of x' was 1). This step is required as otherwise, the norm of the vector will increase without limit at each iteration. With this modification, the next few iterations are:

$$Ax'' = \begin{pmatrix} 2 & -1 \\ -1 & 1 \end{pmatrix}\begin{pmatrix} 1 \\ -\frac{1}{2} \end{pmatrix} = \begin{pmatrix} 5/2 \\ -3/2 \end{pmatrix} = \frac{5}{2}\begin{pmatrix} 1 \\ -3/5 \end{pmatrix}$$

$$\begin{pmatrix} 2 & -1 \\ -1 & 1 \end{pmatrix}\begin{pmatrix} 1 \\ -3/5 \end{pmatrix} = \begin{pmatrix} 13/5 \\ -8/5 \end{pmatrix} = \frac{13}{5}\begin{pmatrix} 1 \\ -8/13 \end{pmatrix}$$

$$\begin{pmatrix} 2 & -1 \\ -1 & 1 \end{pmatrix}\begin{pmatrix} 1 \\ -8/13 \end{pmatrix} = \begin{pmatrix} 34/13 \\ -21/13 \end{pmatrix} = \frac{34}{13}\begin{pmatrix} 1 \\ -21/34 \end{pmatrix}. \quad (5.41)$$

Note that the factor multiplying the vector on the right-hand side of all the above equations (i.e. the component that was 'pulled out') can be considered to be the eigenvalue. Note also that the successive approximations to the eigenvalue are: 2, 2.5, 2.6, 2.61.... One can see that these values are converging to the largest eigenvalue, which, from example 5.4, is seen to be:

$$\frac{3 + \sqrt{5}}{2} = \frac{3 + 2.236}{2} = 2.618.$$

The program can monitor the difference between successive approximations to the eigenvalue and terminate when that difference falls below a preset value (for example, 10^{-3}). If we stop with the last iteration shown in (5.41), the corresponding eigenvector will be $\begin{pmatrix} 1 \\ -21/34 \end{pmatrix}$. Referring back to example 5.4, we see that the components of the eigenvector corresponding to the largest eigenvalue have to satisfy the condition: $\frac{1 - \sqrt{5}}{2} x_1 = x_2 = -0.618$. The power method gives us $x_2 = -\frac{21}{34} = -0.6176$, which is quite close to -0.618.

5.5 Convergence of the power method

The convergence of the power method is based on the assumption that matrix A has distinct eigenvalues and eigenvectors. If the matrix A is a square matrix with a size of $N \times N$, we can write the eigenvalue equations as follows:

$$Au_j = \lambda_j u_j. \tag{5.42}$$

The arbitrary vector x, which is our initial guess for the eigenvector, can be written as the linear combination of all the eigenvectors, i.e.:

$$x = \sum_{j=1}^{N} a_j u_j. \tag{5.43}$$

Let $|\lambda_1| > |\lambda_2| > |\lambda_3| \ldots \ldots |\lambda_N|$; then

$$Ax = A \sum_{j=1}^{N} a_j u_j. \tag{5.44}$$

Since we are repeatedly operating (say m times) on the vector x (5.44), a repeated application of matrix A will be written as:

$$A^m x = A^m \sum_{j=1}^{N} a_j u_j = \sum_{j=1}^{N} a_j \lambda_j^m u_j = \lambda_1^m \left[a_1 u_1 + \sum_{j=2}^{N} a_j \left(\frac{\lambda_j}{\lambda_1} \right)^m u_j \right], \tag{5.45}$$

since $\frac{\lambda_j}{\lambda_1}$ is less than 1 (for all j) $\left(\frac{\lambda_j}{\lambda_1} \right)^m \to 0$ when $m \to \infty$.

Therefore, the last term in (5.45) can be neglected and we get

$$A^m x = \lambda_1^m a_1 u_1. \tag{5.46}$$

Note that 'pulling out' the successive approximations to the eigenvalue at each iteration means that (5.46) can actually be written as:

$$Ax_{m-1} = \lambda_1 a_1 u_1, \tag{5.47}$$

where x_{m-1} is the approximation to the eigenvector at the $(m-1)$th iteration. Note that since the eigenvector is not unique, the factor a_1 in (5.47), which was not there in the eigenvalue equation, should not be an issue.

If, for a given matrix, the largest eigenvalue is not very much greater than the next one, the power method can converge to the next-largest eigenvalue (depending on the starting vector). The power method will not converge to the largest eigenvalue if the starting vector is orthogonal to u_1, i.e. $a_1 = 0$ in (5.45). In this case, the power method will converge to the next-largest eigenvalue instead of the largest eigenvalue.

5.6 Deflation: determination of the remaining eigenvalues

The discussion of the power method in the previous section might not have seemed particularly important to you. After all, you have only managed to determine one of the N eigenvalues of an $N \times N$ matrix! There may well be N eigenvalues, but some are more important than others. It turns out that in some applications (for instance, image compression), it is sufficient to know the first few eigenvalues.

Nevertheless, it turns out that the power method can be modified to determine the remaining eigenvalues as well, but one at a time. After the largest eigenvalue has been found, it is possible to determine the next highest eigenvalue by replacing the original matrix with one that has only the remaining eigenvalues. The process of removing the largest eigenvalue is called deflation. The deflation technique outlined below is designed for symmetric matrices. It exploits the orthogonality of the eigenvectors of symmetric matrices.

Consider the matrix $B = A - \lambda_1 u_1 x^T$, where u_1 is the eigen (column) vector corresponding to the largest eigenvalue and x^T is an arbitrary (row) vector. If the matrix B acts on u_1, we have:

$$Bu_1 = Au_1 - \lambda_1 u_1 x^T u_1 = \lambda_1 u_1 (1 - x^T u_1). \tag{5.48}$$

If x^T is chosen such that $x^T u_1 = 1$, zero is the eigenvalue corresponding to the eigenvector u_1. Deflation changes the largest eigenvalue of A to zero. What about the other eigenvalues? Have they changed due to this transformation? We note that:

$$B^T = A^T - \lambda_1 x u_1^T. \tag{5.49}$$

Let B^T act on any one of the other eigenvectors of A^T. We now have:

$$B^T u_i = A^T u_i - \lambda_1 x u_1^T u_i. \tag{5.50}$$

At this point, $u_1^T u_i = 0$, as the eigenvectors are mutually orthogonal to each other. Therefore, (5.50) reduces to:

$$B^T u_i = A^T u_i; \tag{5.51}$$

thus, the remaining eigenvalues of A (and therefore A^T) are unaffected by the deflation process.

We give below a program that implements the power method to determine the largest eigenvalue and uses deflation to determine the next-largest eigenvalue.

```
'Power method for finding eigenvalues';
clear;
A=[2 -1;-1 1];
 'x is the initial guess for the eigenvector';
for i=1:size(A,1)
    x(i)=1;
end
x=x';
xmax1=1000;
xmax2=1;
iterations=0;
error=abs(xmax1-xmax2);
while error >0.001
    iterations=iterations+1;
    x1=A*x;
    xmax2=max(x1)
    x1=x1/xmax2;
    error=abs(xmax2-xmax1);
    x=x1
    xmax1=xmax2;
end
'deflating the matrix to get the next eigenvalue';
 for i=1:size(A,1)
     v(i)=1/x(i);
 end
 v=v/size(A,1);
 A1=A-xmax1*(x*v);
 for i=1:size(A1,1)
     x2(i)=1;
 end
 x2=x2';
 xmax3=1000;
 xmax4=1;
 iterations=0;
 error=abs(xmax3-xmax4);
 while error >0.5
     iterations=iterations+1;
     x3=A1*x2;
     xmax4=max(x3);
     x3=x3/xmax4;
     error=abs(xmax4-xmax3);
     x2=x3;
     xmax3=xmax4;
 end
```

5.7 Curve fitting: the least-squares technique

Fitting a curve to a set of experimental data points is a common exercise in experimental physics. The reason we are discussing this topic here is that the process followed to fit a curve leads to a system of linear equations, which can be solved using the techniques of linear algebra which we are discussing. We will first discuss an example in which a linear relation is expected between two physical quantities.

For instance, assume that you have measured the electrical resistance of a metallic wire in the laboratory. For many metals, it is expected that the electrical resistance will vary with temperature according to the equation $R = a + bT$. The experimental data you obtained are given below:

T (^{0} C)	19.1	25.0	30.1	36.0	40.0	45.1	50.0
R (Ohm)	76.30	77.80	79.75	80.80	82.35	83.90	85.10

You are now asked to fit these data to the expected linear relation and determine the temperature coefficient of resistance, b. To understand what we mean by this, let us plot the data on a graph sheet.

We note from figure 5.1 that the expected linear relation between the resistance and the temperature is indeed followed, and we can, in fact, draw a straight line that passes through almost all the data points. The slope of such a straight line would give us the temperature coefficient of resistance of the wire,

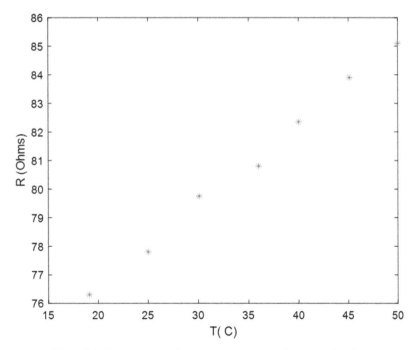

Figure 5.1. Resistance as a function of temperature for a metallic wire.

and its intercept would provide us with the resistance of the wire at 0 °C. However, we note that a single straight line that passes through all the points cannot be drawn. How can we draw the straight line which is most appropriate for the given data points? Should we choose the line that passes through the maximum number of data points?

No. A more appropriate procedure would be to choose the line that minimises the error, where the error is defined by:

$$E = \sum_{i=1}^{N} (y_i - y_{\exp})^2. \tag{5.52}$$

E is nothing more than the sum of the squares of the difference between the expected value of y (in this case, R) for the given x (i.e. temperature T_i) and the actual value of y as measured experimentally. Substituting the expected linear relation into (5.16), we get:

$$E = \sum_{i=1}^{N} (y_i - mx_i - c)^2. \tag{5.53}$$

Since we wish to minimise E, we set its derivatives with respect to m and c to zero (which are the slope intercept of the line which best fits the data). We are differentiating with m and c (instead of x), as we are varying the slope and intercept of the line in order to minimise the error. We cannot vary x, as all the values of x are obtained experimentally and cannot be varied once the experiment has been completed!

Setting the derivatives of E to zero, we arrive at two simultaneous equations:

$$\frac{\partial S}{\partial m} = -2\sum_i x_i(y_i - mx_i - c) = 0 \quad \text{and} \quad \frac{\partial S}{\partial c} = -2\sum_i (y_i - mx_i - c) = 0, \tag{5.54}$$

which give:

$$m\sum x_i^2 + c\sum x_i = \sum x_i y_i \quad \text{and} \quad m\sum x_i + Nc = \sum y_i. \tag{5.55}$$

The second equation can be rewritten as $\bar{y} = m\bar{x} + c$, where $\bar{y} = \frac{1}{N}\sum y_i$ and $\bar{x} = (\frac{1}{N}\sum x_i)$, showing that the best-fit straight line passes through the centroid $(\bar{x}, \bar{y})$ of the points (x_i, y_i). The required values of m and c can be calculated from the above two equations; they are:

$$m = \frac{\sum (x_i - \bar{x})y_i}{\sum (x_i - \bar{x})^2} \quad \text{and} \quad c = \bar{y} - m\bar{x}. \tag{5.56}$$

The best-fit straight line can be drawn by calculating m and c from (5.56). A graphical method of obtaining the best-fit line is to rotate a transparent ruler about the centroid so that it passes through the clusters of points at the top right and at the bottom left. This line will give the maximum error in m, $(\Delta m)_1$ on one side. Do the same to discover the maximum error in m, $(\Delta m)_2$ on the other side. Now bisect the angle between these two lines, and that will be the best-fit line through the experimental data. Note that this graphical procedure is only a crude method to find the best-fit line. It is better to use the expressions given in equation (5.56) to find the slope and intercept of the best-fit line.

5.8 Curve fitting: the generalised least-squares technique

In many cases, the expected relation between the variables is nonlinear. For example, the specific heat of metals has the temperature dependence given in (5.57):

$$c_v = aT^3 + bT. \tag{5.57}$$

Here, the cubic term is the contribution of the lattice to the specific heat, while the linear term is the electronic contribution.

We can use the least-squares technique we have derived if we use the variables c_v/T and T^2 (i.e. we divide (5.57) by T) to fit a straight line with a slope a and intercept b to the given data.

Note that 'linearizing' such a nonlinear expression would not have been possible if a constant term d were present in (5.57).

What do we do in such a case? Fortunately, the least-squares technique is still valid and can be adopted with a suitable modification for such a nonlinear relation.

Assume that a polynomial of degree n fits the data; the sum of the squares of the error would then be:

$$E = \sum_{i=1}^{N}(y_i - y_{\exp})^2, \tag{5.58}$$

where $y_{\exp}$ is a polynomial in x of degree n:

$$y_{\exp} = a_0 + a_1 x + a_2 x^2 + \ldots a_n x^n. \tag{5.59}$$

Substituting $y_{\exp}$ into (5.58) and proceeding along similar lines to (5.54), we set all the derivatives of E to zero to get a system of simultaneous linear equations:

$$\frac{\partial E}{\partial a_0} = -2 \sum_{i=1}^{N}(y_i - a_0 - a_1 x_i - \ldots a_n x_i^n) = 0 \tag{5.60}$$

$$\frac{\partial E}{\partial a_1} = -2 \sum_{i=1}^{N} x_i (y_i - a_0 - a_1 x_i - \ldots a_n x_i^n) = 0$$

. . . .

$$\frac{\partial E}{\partial a_n} = -2\sum_{i=1}^{N} x_i^n(y_i - a_0 - a_1 x - ....a_n x^n) = 0.$$

The (n+1) equations in (5.60) can be rewritten as:

$$a_0 + a_1 \sum_{i=1}^{N} x_i + a_2 \sum_{i=1}^{N} x_i^2 + ...a_n \sum_{i=1}^{N} x_i^n - \sum_{i=1}^{N} y_i = 0 \qquad (5.61)$$

$$a_0 \sum_{i=1}^{N} x_i + a_1 \sum_{i=1}^{N} x_i^2 + a_2 \sum_{i=1}^{N} x_i^3 + ...a_n \sum_{i=1}^{N} x_i^{n+1} - \sum_{i=1}^{N} x_i y_i = 0$$

....

$$a_0 \sum_{i=1}^{N} x_i^n + a_1 \sum_{i=1}^{N} x_i^{n+1} + a_2 \sum_{i=1}^{N} x_i^{n+2} + ...a_n \sum_{i=1}^{N} x_i^{n+1} - \sum_{i=1}^{N} x_i^n y_i = 0.$$

We can use Gaussian elimination to solve these $n+1$ equations, determining the $n+1$ coefficients in the best-fit polynomial. MATLAB has a built-in function, 'polyfit', which fits a polynomial to some given data. This function returns the coefficients of the best-fit polynomial of a specified degree.

In general, this technique of nonlinear regression can be extended to functions that are not polynomials. All you need to do is substitute the expected dependence of y on x into (5.58). The rest of the procedure follows.

Exercises

5.1 Let $A = \begin{bmatrix} 1 \\ 2 \\ 3 \end{bmatrix}$ be the eigenvector corresponding to the largest eigenvalue of a 3×3 matrix. You are trying to determine the eigenvector corresponding to the largest eigenvalue using the power method. Give one choice of initial vector for the power method which will never converge to A (it might converge to the eigenvector corresponding to some other eigenvalue).

5.2 Suppose that you have used the power method to determine the largest eigenvalue λ_1 of a matrix A and v_1 is the corresponding eigenvector. To apply deflation, you compute the matrix $B = A - \lambda_1 v_1 x^T$. Except for the largest eigenvalue of A, the remaining are unaffected by the process of deflation. Will the determinant of B be equal to the determinant of A? If not, what will be the determinant of B be equal to?

5.3

(a) Use row operations to evaluate the following determinant:

$$\begin{vmatrix} 1 & 1 & 1 & \dots & 1 \\ 1 & 1-x & 1 & \dots & 1 \\ 1 & 1 & 2-x & \dots & 1 \\ \dots & \dots & \dots & \dots & \dots \\ 1 & 1 & 1 & \dots & n-x \end{vmatrix}$$

Note that all the off-diagonal elements of the above determinant are equal to one. The final expression which you obtain for the determinant can be a product of many terms (the product need not be simplified further).

(b) How many additions and how many multiplications will be required to evaluate the determinant using the expression you obtained?

(c) Use the expression obtained above to evaluate the value of the determinant if the determinant is of size 4×4 and $x = 5$.

(d) A program has been written below that generates the matrix corresponding to the above determinant:

```
n=10;
for i=1:n
for j=1:n
A(i,j)=1;
end
end.
```

The above part of the program initialises all the elements of the matrix A to one. Write down the remaining part of the program, which will change the appropriate elements of the matrix A to $1-x$, $2-x$, $3-x$, ..., $n-x$. Note: the program is only used to generate the matrix, not to evaluate the determinant.

5.4 Consider the LU decomposition of a matrix A, where L is a lower diagonal matrix with all diagonal elements equal to one. U is an upper diagonal matrix.

(a) Write down the expressions connecting u_{1k} and the elements of the matrix A.

(b) Write down the expressions connecting l_{k1} and the elements of the matrices A and U.

(c) Derive expressions for u_{ik} and l_{kj} in terms of the matrix elements a_{ik} and a_{kj} of the matrix A.

5.5 How many multiplications and how many additions are required to multiply two matrices, each with N rows and N columns? If each multiplication or addition takes one nanosecond, what is the total time required to multiply two matrices that have 100 rows and 100 columns?

5.6 Three blocks which have the same mass $m = 1$ kg are suspended using identical springs whose spring constant is $k = 10$ N m^{-1}, as shown below:

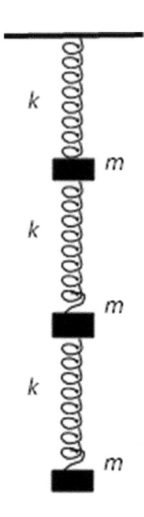

(a) Write down the equations of motion (Newton's second law) for the three masses (ignore gravity, as it does not affect the frequency of oscillation)

(b) Determine the highest normal-mode frequency using the power method (two iterations will be sufficient). Hint: in a normal mode, all the masses will move with the same frequency ω (which is the frequency of that normal mode), but the amplitudes could be different, so the displacement of the ith mass from its equilibrium position is given by $x_i = A_i \exp(i\omega t)$.

5.7 For what kinds of matrices would the LU decomposition procedure outlined in section 5.3 fail? Hint: look at equation (5.40).

5.8 Suppose that you have decomposed a matrix A into the factors L and U using $A = LU$. Do the matrices L and U commute?

5.9 Modify the program in section 5.6 for the power method so that deflation can be repeated to determine all the eigenvalues. Repeated deflation would be required for matrices that have large dimensions. Consider writing the part that performs the deflation as a function that is called repeatedly until all the eigenvalues have been determined.

5.10 Apart from physics, curve fitting and data analysis are used in many areas. The table given below shows the winning time at the 100 m race for men in the Olympic games.

Year	Time (secs)
1896	12.0
1900	11.0
1904	11.0
1908	10.8
1912	10.8
1920	10.8
1924	10.6
1928	10.8
1932	10.3
1936	10.3
1948	10.3
1952	10.4
1956	10.5
1960	10.2
1964	10.0

Year	Time (secs)
1968	9.9
1972	10.14
1976	10.06
1980	10.25
1984	9.99
1988	9.92
1992	9.96
1996	9.84
2000	9.87
2004	9.85
2008	9.69
2012	9.63
2016	9.81
2021	

(a) Plot the points using MATLAB

(b) Determine the best-fit line and draw that line on the graph sheet. Based on your analysis, predict the winning time for the 100 m race in the Olympics to be held in 2021. By the time you read this, the

Olympics have hopefully been held. Compare your 'prediction' with the actual winning time.

(c) Do you think the observed trend[5] (the decrease in the winning time) will continue forever? Why?

5.11 The breaking strength of a certain kind of polyester thread depends on its thickness, as shown in the data given below.

Thickness (mm):	0.1194	0.1778	0.2032	0.2718	0.3150	0.3861	0.4724	0.5867	0.7188	0.8362
Strength (Kg-Wt):	0.7	1.4	3.2	5.0	6.6	9.5	14.1	20.0		44.5

Guess the relation between thickness and strength. Plot the strength on the y-axis and a suitable power of thickness on the x-axis. Determine the parameters of the best-fit line.

If you want to avoid guessing the relation, plot the thickness versus strength graph in MATLAB and make the scale logarithmic on both the x- and y-axes by clicking on the 'property editor' in the 'view' tab of the figure window (where your graph is plotted).

5.12 What are the eigenvalues of the following matrix?

$$\begin{pmatrix} 1 & 0 & 0 & 0 \\ 6 & 2 & 0 & 0 \\ 7 & 4 & 3 & 0 \\ 5 & 6 & 7 & 4 \end{pmatrix}$$

(b) Determine the eigenvectors corresponding to the smallest and largest eigenvalues.

(c) What is the value of the determinant of this matrix?

---

[5] The website https://www.callingbullshit.org/case_studies/case_study_gender_gap_running.html has a humorous, but educative, article on such 'predictions'.

IOP Publishing

Computational Methods Using MATLAB®
An introduction for physicists
P K Thiruvikraman

Chapter 6

Numerical integration and differentiation

In many applications in the physical sciences and engineering, we are required to calculate the derivative of a function that is available only as a set of numbers. This might occur when we have a table of experimentally determined values of some quantity whose derivative is of interest. On other occasions, we need to calculate either a definite integral that cannot be determined by analytical means or one that requires the integral of a function available only as a set of numbers.

We start with a brief discussion of numerical differentiation before passing on to the much more complicated topic of numerical integration, which will take up most of this chapter.

6.1 Numerical differentiation

As we mentioned earlier, imagine that you have a set of values of a function $f(x)$ at some specified values of x, which have been determined experimentally. An expression for the first derivative can be derived from the Taylor series expansion for the function:

$$f(x_2) = f(x_1) + (x_2 - x_1)f'(x_1) \tag{6.1}$$

where we have treated all terms involving higher-order derivatives as though they are negligible.

Equation (6.1) can be rewritten as:

$$f'(x_1) = \frac{f(x_2) - f(x_1)}{(x_2 - x_1)}. \tag{6.2}$$

You might feel that (6.2) is precise (i.e. without error), as it is the definition of the derivative. However, for (6.2) to be the definition of the derivative, (x_2-x_1) must tend to zero, which may not be true if the values of x are taken from a table of experimentally determined values.

doi:10.1088/978-0-7503-3791-5ch6

The error in (6.2) is equal to all the terms in the Taylor series which have been excluded. However, the leading or the most significant part of the error is proportional to the second derivative, which is the next term in the Taylor series. The error in the determination of the derivative obtained using (6.2) is:

$$\frac{(x_2 - x_1)^2}{2!} f''(x_1). \tag{6.3}$$

Recall that each term in the Taylor series is smaller than its predecessor, which is true if $(x_2-x_1) < 1$; therefore, (6.3) is known as the leading term in the error. If we include the term corresponding to the second derivative, then the leading term in the error is reduced. Can we reduce the error? We can try to do this by including one more term in (6.1). This leads to:

$$f(x_2) = f(x_1) + (x_2 - x_1)f'(x_1) + \frac{(x_2 - x_1)^2}{2} f''(x_1). \tag{6.4}$$

We have now introduced the second derivative, which needs to be eliminated. We now evaluate $f(x_0)$ using a Taylor series expansion about $f(x_1)$ to eliminate the second derivative:

$$f(x_0) = f(x_1) + (x_0 - x_1)f'(x_1) + \frac{(x_0 - x_1)^2}{2} f''(x_1). \tag{6.5}$$

Let $(x_2-x_1) = h$ and $(x_0-x_1) = -h$. In other words, we are saying that we have the values of the function at equally spaced intervals, and we wish to calculate its derivative. Subtracting (6.5) from (6.4) eliminates the second derivative. The resulting equation can be rewritten as:

$$f'(x_1) = \frac{f(x_2) - f(x_0)}{(x_2 - x_0)} = \frac{f(x_2) - f(x_0)}{2h}. \tag{6.6}$$

Equation (6.6) is known as the central difference formula, since it approximates the derivative at a point in terms of the difference between the values of the function at two neighbouring points. The error in the central difference formula will be of the order of h^2, as the leading term in the error is now proportional to the third derivative (the second derivative has been eliminated). The error in (6.2), known as the forward difference formula, is proportional to h. Since $h < 1$, a condition required for the convergence of the Taylor series, the error in the central difference formula is less than the error incurred using the forward difference formula. Equation (6.5) is known as the backward difference formula.

Can we reduce the error even more by including further terms in the Taylor series? We will pursue an alternate way of reducing the error in the next section; meanwhile, let us try to arrive at an expression for the second derivative. To derive the finite difference expression for the second derivative, we add (6.4) and (6.5). We then get:

$$f''(x_1) = \frac{f(x_2) + f(x_0) - 2f(x_1)}{h^2}. \tag{6.7}$$

The error in the formula will be proportional to h^4, as the term corresponding to the third derivative is cancelled when we add (6.4) and (6.5).

Example 6.1

Consider the function $f(x) = x\,e^x$. Obtain approximate values for $f'(2)$ with $h = 0.5$ using the finite difference formula for the second derivative.

Applying (6.7) to the given function, we have:

$$f''(2) = \frac{(2 + 0.5)\exp(2.5) + (2 - 0.5)\exp(1.5) - 2\exp(2)}{0.5^2} = 30.49. \qquad (6.8)$$

The actual value of the derivative is:

$$f'(x) = e^x + xe^x \quad \text{and} \quad f''(x) = e^x + e^x + xe^x = 2e^x + xe^x = 29.56. \qquad (6.9)$$

The error introduced by the finite difference formula (6.8) is:

$$\frac{h^2}{12}f''''(x) = \frac{h^2}{12}[4e^x + xe^x]. \qquad (6.10)$$

By substituting $x = 2$ into (6.10), we obtain an error of 0.9236, which is close to the difference between the values obtained in (6.8) and (6.9). Remember that (6.10) only gives the leading term in the error. Therefore, the actual error is the difference between the values in (6.8) and (6.9), which is equal to approximately 0.93.

6.2 The Richardson extrapolation

A technique known as the Richardson extrapolation allows us to obtain a more accurate expression for the derivatives. Including the error term, the central difference formula (6.6), can be written as:

$$f'(x) = \frac{f(x + h) - f(x - h)}{2h} - \frac{h^2}{6}f'''(x). \qquad (6.11)$$

Using a different step size, we can obtain a second approximation for the derivative:

$$f'(x) = \frac{f(x + 2h) - f(x - 2h)}{4h} - \frac{4h^2}{6}f'''(x). \qquad (6.12)$$

In (6.11) and (6.12), we have replaced x_1 with x for the sake of generality. By dividing (6.12) by four and subtracting it from (6.11), we can eliminate the error term that is proportional to h^2. This means that the leading term in the error is proportional to h^4. After performing these operations, we arrive at the following result:

$$\frac{3f'(x)}{4} = \frac{f(x + h) - f(x - h)}{2h} - \frac{f(x + 2h) - f(x - 2h)}{16h} + O(h^4) \qquad (6.13)$$

Or

$$\frac{3f'(x)}{4} = \frac{f(x - 2h) - f(x + 2h) + 8f(x + h) - 8f(x - h)}{16h} + O(h^4). \quad (6.14)$$

Since the derivation of these expressions can be cumbersome, we can perform the Richardson extrapolation numerically instead of symbolically. Let $D_1(h)$ be the approximation to the first derivative obtained using a step size of h and let $D_1(2h)$ be the approximation obtained for the first derivative using a step size of $2h$. A more accurate value for the first derivative is then given by:

$$f'(x) = \frac{4D_1(h) - D_1(2h)}{3}. \quad (6.15)$$

Note that (6.15) is equivalent to (6.14).

6.3 Numerical integration: the area under the curve

We now move on to the more demanding task of numerical integration. As any new student of calculus knows, it is straightforward to differentiate a given function, while the reverse process of integrating a given function and providing the result in terms of elementary functions may or may not be possible.

For example, the integrals $\int_a^b \exp(-x^2)dx$ and $\int_a^b \frac{\sin x}{x}dx$ cannot be evaluated by analytical means when both a and b are finite. Both these integrals can be evaluated by analytical methods if the range of integration is zero to ∞, or $-\infty$ to ∞. We will concentrate our attention on numerical integration while referring the reader to many excellent textbooks[1] which discuss integration by analytical methods in great detail.

The numerical integration of a function of a single variable requires an integral to be evaluated over a certain range, i.e. a definite integral:

$$I = \int_a^b f(x)dx. \quad (6.16)$$

Double or triple integrals can be similarly defined when we integrate a function of many variables over an area or a volume. We will first discuss methods for the integration of a function of a single variable. The definite integral (6.16) has a very simple geometrical interpretation (figure 6.1): I is equal to the area under the curve $f(x)$ from $x = a$ to $x = b$.

Therefore, the definite integral (6.16) is numerically equal to the area enclosed between the curve $f(x)$ and the straight lines $x = a$, $x = b$, and $y = 0$ (the x-axis). If (6.16) can be evaluated by analytical methods, we can do that and substitute the limits to get the numerical value of the integral. If (6.16) cannot be evaluated by analytical methods (using substitution, integration by parts, or conversion into a line

---

[1] The voluminous textbook *Calculus* by G B Thomas, 14th edn, Pearson, London, provides an excellent and comprehensive discussion of all aspects of integration.

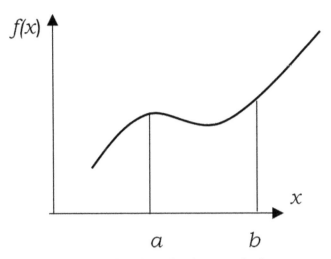

Figure 6.1. The definite integral as the area under the curve.

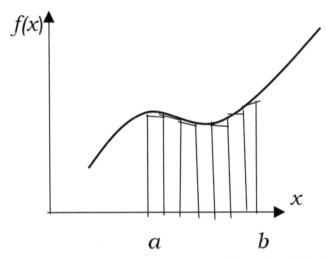

Figure 6.2. Dividing the area under the curve into a finite number of rectangles. Notice the gap between the curve and the tops of the rectangles.

integral in the complex plane), we can evaluate it approximately using numerical methods.

The simplest way to evaluate a definite integral, such as (6.16), is to divide the area under the curve into many rectangles of infinitesimal width and sum the areas of all such rectangles. Figure 6.2 illustrates this procedure.

We notice that dividing the area under the curve into a finite number of rectangles introduces an error into the value we obtain for the integral. Figure 6.2 shows that there are tiny gaps between the tops of the rectangles and the curve $f(x)$. The total error over the entire range of integration is the sum of the errors caused by the individual rectangles.

The procedure illustrated in figure 6.2, known as the rectangular approximation, can be written in the form of an equation as:

$$I = \int_a^b f(x)dx \approx h \sum_{i=0}^{n-1} f(x_i), \tag{6.17}$$

where we have set $a = x_0$ and $b = x_n$ and divided the range of integration into n rectangles. In (6.17), the width of each rectangle is equal to h, where

$$h = \frac{(b-a)}{n}. \tag{6.18}$$

We can use the Taylor series to estimate the error in the rectangular approximation. The function is treated as a constant over the rectangle's width, so the leading term in the error will be due to the term proportional to h in the Taylor series.

$$\int_{x_1}^{x_2} f(x)dx = f(x_1)h + \int_{x_1}^{x_2} (x - x_1)f'(x_1)dx + \dots \tag{6.19}$$

The first term on the right-hand side of (6.19) is the value of the integral obtained using the rectangular approximation, while the second term is the leading term in the error. Without loss of generality, we can take x_1 to be zero in (6.19), which means that the error will be equal to:

$$\frac{h^2}{2}f'(x_1). \tag{6.20}$$

Equation (6.20) shows that the error can be reduced by taking a smaller value of h. Note that at very small values of h, rounding errors become significant. With the available level of precision, we should not use values of h smaller than 10^{-8}. Reducing h would increase the computational load, since we would need to evaluate the function being integrated at a larger number of points. One clever way to reduce the error while avoiding a substantial increase in the computational load is to use the adaptive step-size method. The adaptive step size is based on the fact that the error in the rectangular approximation is proportional to the slope of the function at a point (in addition to its dependence on h). Therefore, the error can be reduced by taking a smaller step size h wherever the slope is steep. When the slope is slight, we switch back to a larger value of h, thereby reducing the computational load.

Note that (6.20) represents the error introduced by evaluating the integral over a range h; to obtain the total error caused by integrating over the interval $[a,b]$, we need to multiply this by n, where n is the number of rectangles. Since $n = (b-a)/h$, the total error in the rectangular approximation is proportional to h. Here, we have ignored the variation in the first derivative of the function over the range of integration.

Equation (6.19) gives us a hint that we can use to improve the rectangular approximation. The second term on the right-hand side of (6.19) corresponds to the error in the rectangular approximation. If we include that term in our approximation to the integral, then the leading term in the error is now the third term in the Taylor series, which is, of course, substantially smaller.

Including one more term in the Taylor series is equivalent to approximating the function by a straight line between two points that differ by the step size h. If we approximate the function by a straight line between two adjacent points, we have trapeziums instead of rectangles; therefore, this approximation is known as the trapezoidal rule.

According to the trapezoidal rule,

$$I = \int_{x_1}^{x_2} f(x)dx \approx \frac{[f(x_1) + f(x_2)]h}{2}. \tag{6.21}$$

The integral over the entire interval [a,b] is equal to the sum of the areas of all the trapeziums; as a result, (6.17) can be replaced by:

$$I = \int_a^b f(x)dx \approx h\left[\frac{f(x_0)}{2} + \sum_{i=1}^{n-1} f(x_i) + \frac{f(x_n)}{2}\right]. \tag{6.22}$$

The error in the trapezoidal rule can be evaluated by noting that we are approximating the function using a straight line. Therefore, the error approximates to the contribution of the quadratic term in the Taylor series. The use of the Taylor series to approximate the function can create a dilemma, as we can expand the function about either x_0 or x_1 (the beginning and end of the interval, respectively). Instead, we note that when we approximate a function by a polynomial of a certain degree, the error can be calculated using (4.10) and (4.14). Applying (4.10) and (4.14) to the current situation, we see that the error over one step size is given by:

$$E = \int_{x_0}^{x_1} \frac{(x - x_0)(x - x_1)}{2!} \left| \frac{d^2f(x)}{dx^2} \right|_{x = \xi} \tag{6.23}$$

$$dx = \int_0^h \frac{x(x - h)}{2!} \left[\frac{d^2f(x)}{dx^2}\right]_{x = \xi} dx = -\frac{h^3}{12}\left[\frac{d^2f(x)}{dx^2}\right]_{x = \xi}.$$

Here, we have taken $x_0 = 0$ and $x_1 = h$.

Equation (6.23) gives the error for one trapezium, the total error over the entire range of integration, i.e. for n trapeziums, is proportional to $h^3n \sim h^2(b-a)$.

6.4 Simpson's rules

We can improve on the trapezoidal rule by approximating the function using a quadratic between the points x_0 to x_2. Deriving the analogue of (6.21) from geometrical considerations becomes quite tedious, so we switch to a new formalism that generally works to approximate the function using a polynomial of arbitrary degree. We note that, in both the trapezoidal rule and the rectangular approximation, we are seeking formulae of the form:

$$I \approx \sum_{i=0}^n w_i f(x_i), \tag{6.24}$$

where the w_i variables represent some weights which have to be determined. If we are approximating the function using a polynomial that passes through two points (as in the trapezoidal rule), the formula (6.24) should give the exact value of the integral for $f(x) = 1$ and $f(x) = x$. In general, if we approximate the function using a polynomial of degree n, (6.24) should give the exact result if the function is a polynomial of degree n or less than n.

If we apply this logic to the case of the trapezoidal rule, we obtain:

$$\int_0^h dx = h = w_0 + w_1 \tag{6.25}$$

and

$$\int_0^h x dx = \frac{h^2}{2} = w_1 h. \tag{6.26}$$

Solving these two simultaneous equations, (6.25) and (6.26), gives us $w_1 = w_0 = h/2$; substituting this into (6.24) gives us the trapezoidal rule (6.21).

If we had used a second-degree polynomial to approximate the function $f(x)$, (6.24) would hold exactly for $f(x) = 1$, $f(x) = x$, and $f(x) = x^2$. Applying these conditions gives us the following three simultaneous equations for the weights.

$$\int_0^{2h} dx = 2h = w_0 + w_1 + w_2 \tag{6.27}$$

$$\int_0^{2h} x dx = 2h^2 = w_1 h + 2w_2 h \tag{6.28}$$

$$\int_0^{2h} x^2 dx = \frac{8h^3}{3} = w_1 h^2 + 4w_2 h^2. \tag{6.29}$$

Equations (6.27)–(6.29) can be solved using the technique of Gaussian elimination discussed in chapter 5. We write the three equations in the form of an augmented matrix:

$$\begin{bmatrix} 1 & 1 & 1 & 2h \\ 0 & 1 & 2 & 2h \\ 0 & 1 & 4 & 8h/3 \end{bmatrix}. \tag{6.30}$$

We subtract the second row from the third row and replace the third row by this difference, i.e. $R_3 - R_2$.

$$\begin{bmatrix} 1 & 1 & 1 & 2h \\ 0 & 1 & 2 & 2h \\ 0 & 0 & 2 & 2h/3 \end{bmatrix}. \tag{6.31}$$

We obtain the weights from (6.31) by back substitution. From the last row of the matrix in (6.31), we can see that $w_2 = h/3$. By substituting this result into the previous row, we get $w_1 = 4h/3$, and by substituting for w_1 and w_2 in the first row, we get $w_0 = h/3$.

Having determined all three weights, we can now substitute them into (6.24) to obtain Simpson's 1/3 rule:

$$\int_{x_0}^{x_2} f(x)dx \approx \frac{h}{3}\big[f(x_0) + 4f(x_1) + f(x_2)\big]. \tag{6.32}$$

In Simpson's 1/3 rule, we approximate the function using a second-degree polynomial; therefore, by an extension of the logic used to obtain (6.23), the error is given by:

$$E = \int_{x_0}^{x_2} \frac{(x - x_0)(x - x_1)(x - x_2)}{3!}\left[\frac{d^3 f(x)}{dx^3}\right]_{x = \xi}$$

$$\tag{6.33}$$

$$dx = \int_0^{2h} \frac{x(x - h)(x - 2h)}{3!}\left[\frac{d^3 f(x)}{dx^3}\right]_{x = \xi} dx = 0.$$

Does (6.33) mean that the error in Simpson's 1/3 rule is zero? Of course not. It just means that we have to proceed to the next term in the infinite series to determine the error:

$$E = \int_{x_0}^{x_2} \frac{(x - x_0)(x - x_1)(x - x_2)(x - x_3)}{4!}\left[\frac{d^4 f(x)}{dx^4}\right]_{x = \xi}$$

$$\tag{6.34}$$

$$dx = \int_0^{2h} \frac{x(x - h)(x - 2h)(x - 3h)}{4!}\left[\frac{d^4 f(x)}{dx^4}\right]_{x = \xi} dx = -\frac{h^5}{90}\left[\frac{d^4 f(x)}{dx^4}\right]_{x = \xi}.$$

Equation (6.34) gives the error between x_0 and x_2; to obtain the total error over the entire range of integration, we would have to multiply this result by $n/2$, so the total error is proportional to h^4.

We note from (6.34) that Simpson's 1/3 rule has an error that is an order of magnitude smaller than what was expected based on the degree of the polynomial used to approximate the function. This is the reason why this formula is quite popular.

We can improve on Simpson's 1/3 rule by approximating the function using a third-degree polynomial that passes through the four points, x_0, x_1, x_2, and x_3. Therefore, (6.24) would hold exactly for $f(x) = 1$, $f(x) = x$, $f(x) = x^2$, and $f(x) = x^3$. Applying these conditions gives us the following four simultaneous equations for the weights:

$$\int_0^{3h} dx = 3h = w_0 + w_1 + w_2 + w_3 \tag{6.35}$$

$$\int_0^{3h} xdx = \frac{9h^2}{2} = w_1 h + 2w_2 h + 3w_3 h \tag{6.36}$$

$$\int_0^{3h} x^2 dx = 9h^3 = w_1 h^2 + 4w_2 h^2 + 9w_3 h^2 \tag{6.37}$$

$$\int_0^{3h} x^3 dx = \frac{81h^4}{4} = w_1 h^3 + 8w_2 h^3 + 27w_3 h^3. \tag{6.38}$$

Equations (6.35)–(6.38) can be solved using the technique of Gaussian elimination discussed in chapter 5. We can write these four equations in the form of an augmented matrix:

$$\begin{bmatrix} 1 & 1 & 1 & 1 & 3h \\ 0 & 1 & 2 & 3 & 9h/2 \\ 0 & 1 & 4 & 9 & 9h \\ 0 & 1 & 8 & 27 & 81h/4 \end{bmatrix}. \tag{6.39}$$

We subtract the second row from the third row and replace the third row by this difference, i.e. $R_3 - R_2$. Similarly, we subtract the second row from the fourth row and replace the fourth row by the difference to obtain:

$$\begin{bmatrix} 1 & 1 & 1 & 1 & 3h \\ 0 & 1 & 2 & 3 & 9h/2 \\ 0 & 0 & 2 & 6 & 9h/2 \\ 0 & 0 & 6 & 24 & 63h/4 \end{bmatrix}. \tag{6.40}$$

We now subtract the third row from the last row three times to get the augmented matrix in reduced echelon form:

$$\begin{bmatrix} 1 & 1 & 1 & 1 & 3h \\ 0 & 1 & 2 & 3 & 9h/2 \\ 0 & 0 & 2 & 6 & 9h/2 \\ 0 & 0 & 0 & 6 & 9h/4 \end{bmatrix}. \tag{6.41}$$

We obtain the weights from (6.41) by back substitution. From the last row of the matrix in (6.41), we can see that: $w_3 = 3h/8$. By substituting this result into the previous row, we get $w_2 = 9h/8$, and by substituting for w_3 and w_2 in the second row, we get $w_1 = 9h/8$. Finally, we also get $w_0 = 3h/8$.

Having determined all four weights, we can now substitute them into (6.24) to obtain Simpson's 3/8 rule:

$$\int_{x_0}^{x_3} f(x)dx \approx \frac{3h}{8}\left[f(x_0) + 3f(x_1) + 3f(x_2) + f(x_3)\right]. \tag{6.42}$$

Note that in both Simpson's 1/3 rule and the 3/8 rule, the middle point(s) are given more weight, which is intuitively obvious. The error in Simpson's 3/8 rule is lower than the error in the 1/3 rule, but it is of the same order.

In principle, we can continue this process of approximating the function by a polynomial of a certain degree and derive a five-point formula or a six-point formula (see problem 6.6 at the end of this chapter). Still, in practice, this is rarely done, as the derivation of rules with more points becomes more and more cumbersome, and a more accurate method exists that uses fewer points. We will discuss this superior method, known as Gaussian quadrature. Before we proceed to a discussion of Gaussian quadrature, let us compare all the quadrature methods discussed so far and compare the amount of computation required by these methods.

6.5 Comparison of quadrature methods

While we have calculated the error involved in each of the quadrature methods discussed so far, we have not compared the amount of computation involved in these methods. We now examine this aspect more closely.

The computation involved in quadrature methods involves the following operations:

    (i) Function evaluations
    (ii) Additions
    (iii) Multiplications and divisions

If we fix a certain step size for the range of integration ($x = a$ to $x = b$), we have fixed the number of points and hence, the number of function evaluations. We will use n to denote the number of points involved. We note from (6.17) that the number of function evaluations is n and that there are $n-1$ additions and one multiplication involved in the rectangular approximation. From (6.22), we see that the composite formula for the trapezoidal rule involves $n+1$ function evaluations, n additions, one division (by two) and one multiplication (by h). The composite formula for Simpson's 1/3 rule can be written as:

$$\int_{x_0}^{x_n} f(x)dx \approx \frac{h}{3}\left[f(x_0) + f(x_n) + 4\sum_{i=1}^{\frac{n}{2}-1} f(x_{2i+1}) + 2\sum_{i=1}^{\frac{n}{2}-1} f(x_{2i}) \right]. \tag{6.43}$$

Here, it has been assumed that n is an even number. We see from (6.43) that we require $n-1$ additions, three multiplications, and one division (by three) to evaluate the integral using Simpson's 1/3 rule.

From the preceding discussion, we see that there is a marginal increase in the computational load when we proceed from the rectangular approximation to the Simpson's rule, but that is a small price to pay for the tremendous increase in accuracy that results.

6.6 Romberg integration

As in the case of numerical differentiation, we can improve the accuracy of the quadrature schemes we are using by determining the value of the integral for two different step sizes (values of h). These two values for the integral, which correspond to two different step sizes, can then be used to arrive at an even more accurate estimate for the integral. The procedure is exactly same as that followed for Richardson extrapolation.

Suppose we are using the trapezoidal rule; the total error over the entire range of integration is proportional to h^2. Let I_e be the exact value of an integral and I_h and $I_{h/2}$ be the values determined using step sizes of h and $h/2$, respectively.

We have:

$$I_e = I_h + kh^2$$

and

$$I_e = I_{h/2} + \frac{kh^2}{4}.$$ (6.44)

We can use the two equations in (6.44) to determine k. Substituting this value of k into either of the equations helps us to determine the error up to the order of h^2. In other words, the order of the error is now proportional to a higher power of h. Therefore, the error is reduced.

6.7 Gaussian quadrature

The legendary mathematician Gauss, who contributed to many fields in mathematics and physics, observed that the quadrature methods could yield more accurate results if we use uneven spacing for the x values. Gauss observed that if we remove the even spacing, a three-term quadrature formula would contain six parameters (three unknown values of x and three weights). This corresponds to a fifth-degree interpolation formula. Since the degree of the polynomial has increased, the error will also be reduced.

Consider a two-term formula. There are four unknowns; therefore, this corresponds to a third-degree polynomial (as far as accuracy is concerned). Therefore, this formula should be exact for $f(x) = 1$, $f(x) = x$, $f(x) = x^2$, and $f(x) = x^3$.

In general, for an n-point formula, we have:

$$I = \int_a^b f(x)dx \approx \sum_{i=1}^n w_i f(x_i).$$ (6.45)

We need to determine the $2n$ unknown parameters (n weights and n values of x). The Gaussian quadrature formula (6.45) will give exact results if $f(x)$ is a polynomial of order $2n-1$ or less. We use this fact to determine the weights and the abscissae (values of x).

Gauss came up with the following integral to determine the weights and the abscissae:

$$\int_a^b q_{n-1}(x)\varphi_n(x)dx = \sum_{i=1}^n w_i q_{n-1}(x_i)\varphi_n(x_i).$$ (6.46)

Here, $q_{n-1}(x)$ is an arbitrary polynomial of degree $n-1$ and $\phi_n(x)$ is an orthogonal polynomial over the range a to b. For integration over a finite range, it helps to use the Legendre polynomials, which are orthogonal polynomials defined over the range $x = -1$ to $x = +1$. If the range of integration (a,b) is different from $(-1,1)$, we can apply a transformation to change the range to $(-1,1)$. In fact, any range, whether finite or infinite, can be transformed to $(-1,1)$ by applying a suitable transformation.

The set of orthogonal polynomials $\phi_i(x)$ (i=1 to n) satisfy the orthogonality relation:

$$\int_a^b \varphi_l(x)\varphi_m(x)dx = \delta_{lm}.$$ (6.47)

We have assumed in (6.47) that the polynomials $\phi_i(x)$ (i=1 to n) are orthonormal. If the polynomials are not normalised, we have to modify the right-hand side of (6.47).

A set of orthogonal polynomials will form a complete set of basis functions, and therefore, the arbitrary polynomial $q_{n-1}(x)$ can be expanded in terms of the orthogonal polynomials as:

$$q_{n-1}(x) = \sum_{i=0}^{n-1} q_i \varphi_i(x). \tag{6.48}$$

The q_i in (6.48) are the expansion coefficients. Substituting (6.48) into (6.46), we find:

$$\sum_{i=0}^{n-1} \int_a^b q_i \varphi_i(x) \varphi_n(x) dx = \sum_{i=1}^{n} w_i q_{n-1}(x_i) \varphi_n(x_i). \tag{6.49}$$

Using the orthogonality relation (6.47), we find that the left-hand side of (6.49) is zero. Since the q_i values are arbitrary, the right-hand side is zero if the abscissa x_i are the zeros of the orthogonal polynomial. In fact, x_i have to be the zeros of the orthogonal polynomials, as otherwise, (6.49) will not hold.

To determine the weights corresponding to each x_i, we can use the Lagrange interpolation polynomial we saw in chapter 4. Since (6.45) is exact for polynomials of degrees up to $2n-1$, we can use a Lagrange interpolation of degree $n-1$ to determine the weights.

The Lagrange interpolation polynomial of degree $n-1$ is given by:

$$l_{j,n}(x) = \frac{(x - x_1)(x - x_2)..(x - x_{j-1})(x - x_{j+1})..(x - x_n)}{(x_j - x_1)(x_j - x_2)...(x_j - x_{j-1})(x_j - x_{j+1})...(x_j - x_n)}. \tag{6.50}$$

It has the interesting property that:
$l_{j,n}(x_i) = 0$ if $j \neq i$
$l_{j,n}(x_i) = 1$, if $j = i$.
Therefore:

$$\int_a^b l_{j,n}(x) dx = \sum_{i=1}^{n} w_i l_{j,n}(x_i) = w_j. \tag{6.51}$$

Therefore, the weights can be obtained by analytically performing the above integration.

Let us now see how we can use the concepts discussed above to develop a two-point Gaussian quadrature formula. We have to use the zeros of the Legendre second-degree polynomial as the abscissae of the two-point formula.

The Legendre second-degree polynomial is:

$$\varphi_2(x) = \frac{3x^2 - 1}{2} \tag{6.52}$$

The zeros of this polynomial occur at $\pm\sqrt{\frac{1}{3}}$. The corresponding weights are:

$$w_1 = \int_{-1}^{1} \frac{x - x_2}{x_1 - x_2} dx = -\frac{\sqrt{3}}{2} \left[\int_{-1}^{1} \left(x - \frac{1}{\sqrt{3}} \right) dx \right] = 1 \tag{6.53}$$

$$w_2 = \int_{-1}^{1} \frac{x - x_1}{x_2 - x_1} dx = \frac{\sqrt{3}}{2} \left[\int_{-1}^{1} \left(x + \frac{1}{\sqrt{3}} \right) dx \right] = 1. \tag{6.54}$$

To derive the Gaussian quadrature formula for $n=3$ (a three-point formula), we use the Legendre third-degree polynomial:

$$\varphi_3(x) = \frac{5x^3 - 3x}{2}. \tag{6.55}$$

The zeros of this polynomial are at $x = -\sqrt{\frac{3}{5}}$, 0, and $\sqrt{\frac{3}{5}}$. The corresponding weights are:

$$w_1 = \int_{-1}^{1} \frac{(x - x_2)(x - x_3)}{(x_1 - x_2)(x_1 - x_3)} dx = \frac{5}{6} \left[\int_{-1}^{1} x \left(x - \sqrt{\frac{3}{5}} \right) dx \right] = \frac{5}{9} \tag{6.56}$$

$$w_2 = \int_{-1}^{1} \frac{(x - x_1)(x - x_3)}{(x_2 - x_1)(x_2 - x_3)} dx = \frac{-5}{3} \left[\int_{-1}^{1} \left(x + \sqrt{\frac{3}{5}} \right) \left(x - \sqrt{\frac{3}{5}} \right) dx \right] = \frac{8}{9} \tag{6.57}$$

$$w_3 = \int_{-1}^{1} \frac{(x - x_1)(x - x_2)}{(x_3 - x_1)(x_3 - x_2)} dx = \frac{5}{6} \left[\int_{-1}^{1} x \left(x + \sqrt{\frac{3}{5}} \right) dx \right] = \frac{5}{9}. \tag{6.58}$$

The evaluation of the weights and zeros for higher-order Legendre polynomials involves a significant amount of calculation, but the labour involved can be reduced by noting the following:

(i) For Legendre polynomials of odd degree, zero is always one of the roots.

(ii) The zeros of Legendre polynomials of odd or even degrees are always symmetric about the origin. This follows from the fact that the Legendre polynomials are polynomials in $\cos(\theta)$, which is an even function[2].

(iii) The weights for two zeros symmetrically placed with respect to the origin are always equal.

(iv) When we evaluate the weights, we can immediately set all integrals of odd functions to zero, since the integral is from -1 to 1 (the limits are symmetric with respect to the origin).

(v) The sum of all the weights will be equal to two.

While the above rules help us to avoid unnecessary calculations, we present some more rules which help us to detect any errors we may make during our calculations:

i. The weight for a zero closer to the origin is greater than that for a zero which is farther away.

ii. The weight for the origin (if the origin is one of the zeros) will be the highest.

---

[2] *Introduction to Electrodynamics* by D J Griffiths, Pearson, London, 1999, provides an excellent introduction to the connection between Legendre polynomials and electrodynamics. *A First Course in Computational Physics* by Paul DeVries and Javier Hasbun, 2nd edition, Jones & Bartlett, Burlington, MA, 2011, provides a thorough discussion of Gaussian quadrature and tabulates the weights and zeros.

Example 6.2

Evaluate $\int_{-1}^{1} x^6 dx$ using the two-term Gaussian quadrature formula.

Using the weights and zeros given in (6.52)–(6.54), we have:

$$\int_{-1}^{1} x^6 dx \approx \left(-\frac{1}{\sqrt{3}}\right)^6 + \left(\frac{1}{\sqrt{3}}\right)^6 = \frac{2}{27}. \qquad (6.59)$$

If we had used the trapezoidal rule with only two points, we would have obtained:

$$\int_{-1}^{1} x^6 dx \approx \frac{1}{2}\left[(-1)^6 + (1)^6\right] = 1. \qquad (6.60)$$

The exact value of the integral is, of course,

$$\int_{-1}^{1} x^6 dx = \left[\frac{x^7}{7}\right]_{-1}^{1} = \frac{2}{7}. \qquad (6.61)$$

It is interesting to observe how the value given by Gaussian quadrature approaches the exact value as we increase the order of the Legendre polynomial (i.e. the number of points used).

For $n = 3$, we have:

$$\int_{-1}^{1} x^6 dx \approx \frac{5}{9}\left(-\sqrt{\frac{3}{5}}\right)^6 + \frac{8}{9}(0)^6 + \frac{5}{9}\left(\sqrt{\frac{3}{5}}\right)^6 = \frac{6}{25}. \qquad (6.62)$$

For $n \geqslant 4$, the error in the Gaussian quadrature reaches zero for this integral and we get the exact result.

6.8 Gaussian quadrature for arbitrary limits

If the limits are from a to b (where $a \neq -1$ and $b \neq 1$), we have to scale the interval to map it to the interval $[-1,1]$. Let the original range for the integration over x be $[a,b]$. We now transform the integral to one over the variable t, such that the range of integration is from $t = -1$ to $+1$. We have to arrange the relation between x and t so that the interval [a,b] is mapped to the interval $[-1,1]$. For the interval [a,b] for x to be mapped to the interval $[-1,1]$ for t, we must have:

$$\frac{x - a}{t + 1} = \frac{b - a}{2}. \qquad (6.63)$$

Rearranging the terms in (6.63), we have:

$$x = a + \frac{(t + 1)(b - a)}{2}. \qquad (6.64)$$

Differentiating (6.64), we get:

$$dx = \frac{dt(b - a)}{2}.$$ (6.65)

Using (6.64) and (6.65), we can transform the integral over x to one over t:

$$\int_a^b f(x)dx = \frac{(b - a)}{2} \int_{-1}^1 f\left[a + \frac{(t + 1)(b - a)}{2}\right]dt$$ (6.66)

The following example illustrates the procedure used to apply (6.66).

Example 6.3

Evaluate $\int_0^{\pi/2} \sin x dx$ using a two-term Gaussian quadrature.

Since the range of integration is from $[0,\pi/2]$ instead of $[-1,1]$, we apply the change of variables (6.64). The given integral then transforms to:

$$\int_0^{\pi/2} \sin x dx = \frac{\pi}{4} \int_{-1}^1 \sin\left[\frac{(t + 1)\pi}{4}\right]dt.$$ (6.67)

Using the weights and zeros in (6.51) to (6.53), we can evaluate the integral over t in (6.67):

$$\frac{\pi}{4} \int_{-1}^1 \sin\left[\frac{(t + 1)\pi}{4}\right]dt = \frac{\pi}{4}\left[\sin\left[\frac{\pi}{4}\left(-\frac{1}{\sqrt{3}} + 1\right)\right] + \sin\left[\frac{\pi}{4}\left(\frac{1}{\sqrt{3}} + 1\right)\right]\right] = 0.99847.$$ (6.68)

Using the trapezoidal rule, which also uses two points, we would have obtained:

$$\int_0^{\pi/2} \sin x dx = \frac{\pi}{4} \sin\left(\frac{\pi}{2}\right) = 0.7854.$$ (6.69)

The exact value of the integral is one. Using just two function evaluations, the Gaussian quadrature method obtains a result that is within 0.2% of the exact result! By comparison, the trapezoidal rule has an error of more than 21% for the same number of function evaluations.

The transformation of variables adopted here is also useful when one (or both) of the limits is (are) ∞.

We routinely come across integrals of the form $\int_0^\infty f(x)dx$. One way to evaluate such integrals is to split them in two, i.e.:

$$\int_0^\infty f(x)dx = \int_0^a f(x)dx + \int_a^\infty f(x)dx$$ (6.70)

The two integrals in (6.70) can be evaluated individually using Gaussian quadrature, but we have to first deal with the second integral. Since the problem is with the second integral in (6.70), we apply the following change of variables.

Let $x = 1/y$.

Then

$$\int_a^\infty f(x)dx = -\int_{1/a}^0 f\left(\frac{1}{y}\right)\frac{dy}{y^2}. \tag{6.71}$$

The application of (6.71) is clear from the following example.

Example 6.4

Evaluate the integral $\int_0^\infty \frac{dx}{1+x^2}$ using the two-point Gaussian quadrature formula.

This integral can easily be evaluated by analytical means, but we wish to illustrate the procedure for converting the range of integration over a finite range. This problem also provides us with another opportunity to calculate the error in the Gaussian quadrature, as the exact value of the integral is known.

As explained earlier, we split the interval $[0, \infty]$ into two parts: $[0, a]$ and $[a, \infty]$; a can be taken to be one. Therefore:

$$\int_0^\infty \frac{dx}{1+x^2} = \int_0^1 \frac{dx}{1+x^2} + \int_1^\infty \frac{dx}{1+x^2}. \tag{6.72}$$

The first integral can be evaluated using Gaussian quadrature, but the second integral needs to be transformed. Let $x = 1/y$ in the second integral only.

$$\int_1^\infty \frac{dx}{1+x^2} = \int_1^0 \frac{-dy}{y^2\left(1+\dfrac{1}{y^2}\right)} = \int_0^1 \frac{dy}{1+y^2} \tag{6.73}$$

Therefore, both the integrals on the right-hand side of (6.72) are identical, i.e. the contribution to the integral made by the interval $[0,1]$ is equal to the contribution made by the interval $[1,\infty]$. This is because the integrand decreases rapidly beyond $x = 1$. Therefore, the given integral reduces to:

$$\int_0^\infty \frac{dx}{1+x^2} = 2\int_0^1 \frac{dx}{1+x^2}. \tag{6.74}$$

To evaluate the integral on the right-hand side of (6.74), we need to transform the range to $[-1,1]$ so that we can use Gaussian quadrature. The required transformation is:

$$\frac{x-0}{t+1} = \frac{1}{2} \tag{6.75}$$

Using (6.75) on the right-hand side of (6.74), we have:

$$2\int_0^1 \frac{dx}{1+x^2} = \int_{-1}^1 \frac{dt}{1+\left(\dfrac{t+1}{2}\right)^2}. \tag{6.76}$$

Using the two-point Gaussian quadrature, we have:

$$\int_{-1}^{1} \frac{dt}{1 + \left(\dfrac{t+1}{2}\right)^2} = \frac{1}{1 + \dfrac{1}{4}\left(1 + \dfrac{1}{\sqrt{3}}\right)^2} + \frac{1}{1 + \dfrac{1}{4}\left(1 - \dfrac{1}{\sqrt{3}}\right)^2} = 1.5738. \tag{6.77}$$

The exact analytical result is:

$$\int_{0}^{\infty} \frac{dx}{1 + x^2} = tan^{-1}(x)_0^\infty = 1.5708. \tag{6.78}$$

Once again, the Gaussian quadrature method has proven itself to be incredibly accurate! It seems to be good to be true. As the following example illustrates, Gaussian quadrature (as well all schemes for numerical integration) can go terribly wrong.

Example 6.5

Determine the transformation which converts the interval $[-\infty,\infty]$ into the interval $[-1,1]$ and use the transformation to perform a numerical evaluation of the integral $\int_{-\infty}^{\infty} \left[\dfrac{1 - \exp(-2x)}{1 + \exp(-2x)}\right]^2 dx$. Use a two-term Gaussian quadrature to evaluate the transformed integral over the interval $[-1, 1]$. Do you think the numerical answer you have obtained is a reasonable estimate of the integral? Why?

Answer: The transformation $y = \tanh x = \dfrac{\exp(x) - \exp(-x)}{\exp(x) + \exp(-x)}$ maps the interval $[-\infty,\infty]$ to the interval $[-1,1]$.

By applying this transformation to the integral given to us, we obtain:

$$y = \frac{\exp(x) - \exp(-x)}{\exp(x) + \exp(-x)} \tag{6.79}$$

$$dy = 1 - \left[\frac{\exp(x) - \exp(-x)}{\exp(x) + \exp(-x)}\right]^2 = (1 - y^2)dx \tag{6.80}$$

or $\dfrac{dy}{(1 - y^2)} = dx$.

Substituting this back into the integral and applying the transformation $y = \tanh$ (x), we obtain $\int_{-1}^{1} \dfrac{y^2}{1 - y^2} dy$. Using a two-term Gaussian quadrature formula, the numerical value of the given integral is

$$\frac{\left(\dfrac{-1}{\sqrt{3}}\right)^2}{1 - \left(\dfrac{-1}{\sqrt{3}}\right)^2} + \frac{\left(\dfrac{1}{\sqrt{3}}\right)^2}{1 - \left(\dfrac{1}{\sqrt{3}}\right)^2} = 1. \tag{6.81}$$

The actual value of the given integral is ∞ (the integral diverges). This can be seen from the fact that the integrand is positive over the whole domain (for both negative and positive values of x) and it increases quickly to the value +1 within a short distance from the origin.

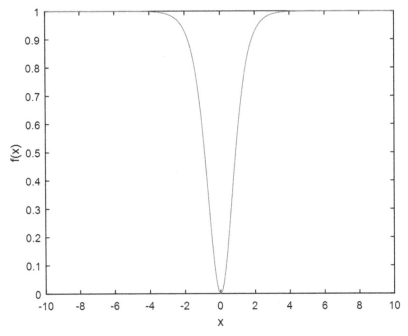

Figure 6.3. Plot of the integrand of example 6.5.

Therefore, Gaussian quadrature or any other numerical method will give an incorrect answer (figure 6.3). The fact that the integral is ∞ can also be seen by plotting the integrand.

6.9 Improper integrals

Example 6.5 should have convinced you that numerical methods have to be used with caution. If you use numerical methods blindly, you are likely to end up with a numerical result that is totally wrong. Analytical methods should always be used to check the correctness of numerical results.

The integral in example 6.5 is an example of an improper integral. A definite integral is called improper if:

(i) One or both of the limits of integration is/are infinite. Such integrals are known as type-I improper integrals.

(ii) The integrand is discontinuous somewhere in the domain of integration, [a,b]. Such integrals are known as type-II improper integrals.

The convergence of improper integrals can be tested using (i) a limit comparison test or (ii) a direct comparison test.

6.9.1 Limit comparison test

We wish to numerically evaluate the integral $\int_a^\infty f(x)dx$. Let us say that we cannot evaluate this integral analytically, but we can analytically evaluate another integral, $\int_a^\infty g(x)dx$. The limit comparison test then states that if $lt\frac{f(x)}{g(x)} = L$ as x tends to infinity, where L is finite, the integrals $\int_a^\infty f(x)dx$ and $\int_a^\infty g(x)dx$ will both converge, or both will diverge. Since we are sure about the convergence of $\int_a^\infty g(x)dx$, we now can be sure about the convergence of $\int_a^\infty f(x)dx$.

Example 6.6

Determine whether the following integral converges by applying the limit comparison test.

$$\int_1^\infty \frac{1}{x}\sqrt{1 + \frac{1}{x^4}}\, dx$$

We can compare the integrand with $1/x$.
As $x \to \infty$, we have

$$\frac{\frac{1}{x}\sqrt{1 + \frac{1}{x^4}}}{\frac{1}{x}} = \sqrt{1 + \frac{1}{x^4}} = 1. \tag{6.82}$$

Therefore, the integrals $\int_1^\infty \frac{1}{x}\sqrt{1 + \frac{1}{x^4}}\, dx$ and $\int_1^\infty \frac{1}{x}dx$ will both converge or both will diverge, since their ratio of the integrands tends to one as $x \to \infty$. But we know that $\int_1^\infty \frac{1}{x}dx$ diverges, as $\int_1^\infty \frac{1}{x}dx = \ln(x)|_1^\infty = \ln\infty = \infty$. Therefore, the given integral also diverges.

6.9.2 Direct comparison test

In the direct comparison test, if we can show that $\int_a^\infty f(x)dx < \int_a^\infty g(x)dx$ and if $\int_a^\infty g(x)dx$ converges, then $\int_a^\infty f(x)dx$ will surely also converge. Showing that $\int_a^\infty f(x)dx < \int_a^\infty g(x)dx$ is easy if both the integrands are positive throughout the range of integration. To show that one integral is less than the other, we then only need to show that $f(x) < g(x)$, as the following example demonstrates.

Example 6.7

Show that $\int_1^\infty \frac{dx}{\exp(x) + 1}$ converges. Integrals of this type occur when we are dealing with the Fermi–Dirac distribution.

Answer: we know that $\frac{1}{\exp(x) + 1} < \frac{1}{\exp(x)}$ throughout the range of integration. Therefore:

$$\int_1^\infty \frac{dx}{\exp(x) + 1} < \int_1^\infty \frac{dx}{\exp(x)} = -\exp(-x)\,|_1^\infty = \exp(-1). \tag{6.83}$$

Equation (6.83) proves the convergence of $\int_1^\infty \frac{dx}{\exp(x) + 1}$ and also gives us an upper bound on the value of the integral. An upper bound is useful. If we get a numerical value (using some method) that is greater than the upper bound, then we surely need to verify our numerical scheme or its implementation.

Note that if $\int_a^\infty f(x)dx > \int_a^\infty g(x)dx$ and $\int_a^\infty g(x)dx$ converges, we cannot come to any conclusion regarding the convergence of $\int_a^\infty f(x)dx$. However, if $\int_a^\infty f(x)dx > \int_a^\infty g(x)dx$ and $\int_a^\infty g(x)dx$ is known to diverge, then clearly $\int_a^\infty f(x)dx$ also diverges.

The efficacy of the direct comparison test depends on our ability to come up with a suitable $g(x)$. We present one more example to stress this vital point.

Example 6.8

Does the integral $\int_2^\infty \frac{dx}{\ln(x)}$ converge or diverge? Use the direct comparison test to decide on the convergence.

Over the entire range of integration, $\ln(x) < x$. Therefore:

$$\frac{1}{\ln(x)} > \frac{1}{x} \tag{6.84}$$

$$\int_2^\infty \frac{dx}{\ln(x)} > \int_2^\infty \frac{dx}{x} = \ln(x)\big|_2^\infty = \infty. \tag{6.85}$$

Therefore, we see from (6.85) that the given integral diverges.

So far, we have only discussed improper integrals of type I. Even if a type-II improper integral converges, we still need to deal with it carefully. The integrand of a type-II integral may diverge at a point within the domain of integration. The integral could still converge to a finite value. However, you may get an incorrect value if the function is evaluated at the point at which the function diverges. If a function diverges at a point and you try to evaluate its value at that point, MATLAB will return 'NaN' (not a number). As a result, the value of the integral will also be returned as 'NaN'.

To avoid this kind of error, we can choose the points so as to avoid points for which MATLAB returns 'NaN'. For instance, if you wish to evaluate $\int_0^{2\pi} \frac{\sin(x)}{x}dx$, then the point $x = 0$ should be avoided, as MATLAB will return NaN at that point. Note that the limit of this function exists at $x = 0$ (the function is equal to one). You can easily evaluate the limit of such functions at many points and confirm that the limit exists, but then MATLAB does not know series expansions or L'Hospital's rule (at least, not yet!).

Sometimes, there may be a singularity at a point within the domain of integration, but we can remove the singularity (and hence evaluate the integral) by a change of variables, as the following example shows.

Example 6.9

Discuss the convergence of $\int_0^1 \frac{\cos(x)}{\sqrt{x}}dx$.

We only expect a problem at $x = 0$, as the integrand apparently diverges at that point. Let us confirm this by trying to evaluate the limit of the integrand as $x \to 0$.

$$\lim_{x \to 0} \frac{\cos x}{\sqrt{x}} = \frac{\left(1 - \frac{x^2}{2} + \frac{x^4}{4!} - ..\right)}{\sqrt{x}} = x^{-1/2} - \frac{x^{3/2}}{2} + \frac{x^{7/2}}{24} - .. \qquad (6.86)$$

Clearly, the integrand diverges as $x \to 0$. By changing the variables, i.e. $x = u^2$, we can remove the singularity:

$$\int_0^1 \frac{\cos(x)}{\sqrt{x}} dx = \int_0^1 \frac{\cos(u)^2}{u} 2u\, du = 2 \int_0^1 \cos(u)^2 du. \qquad (6.87)$$

The change of variables in (6.87) has removed the singularity, and we can now evaluate the value of the transformed integral by any of the quadrature methods.

The singularity can also be removed by subtracting it away. For instance, consider the integral $\int_0^1 \frac{dx}{(1+x)\sqrt{x}}$. The integrand has a singularity at $x = 0$. We can eliminate the singularity by rewriting the given integral as:

$$\int_0^1 \frac{dx}{(1+x)\sqrt{x}} = \int_0^1 \left[\frac{1}{\sqrt{x}} + \frac{1}{(1+x)\sqrt{x}} - \frac{1}{\sqrt{x}} \right] dx. \qquad (6.88)$$

The modified integral in (6.88) simplifies to:

$$\int_0^1 \frac{dx}{\sqrt{x}} + \int_0^1 \frac{[1-(1+x)]dx}{(1+x)\sqrt{x}} = 2\sqrt{x} \Big|_0^1 - \int_0^1 \frac{\sqrt{x}\, dx}{1+x}. \qquad (6.89)$$

The integrals in (6.89) do not have a singularity at $x = 0$.

We can also remove the singularity by multiplying and dividing the integrand by a suitable term.

Sometimes, we encounter problems even if an integral does not have a singularity. Consider the integral $\int_0^1 x^{2/3} dx$. Even though there is no singularity in the integral, a straightforward evaluation using a quadrature method will give rise to a large error in the result. The error is greater for this particular integral because the derivatives of the integrand become unbounded at $x = 0$. Recall that the error in the quadrature methods depends on the derivative of the integrand (sections 6.3 and 6.4). We can reduce the error by changing the variables as follows: $x = u^3$. This change of variables transforms the integral, and we obtain:

$$\int_0^1 x^{2/3} dx = \int_0^1 3u^2 u^2 du = \int_0^1 3u^4 du.$$

The derivatives of the integrand do not have any singularity and therefore the error is also considerably reduced.

6.10 Approximate evaluation of integrals using Taylor series expansion

Consider the integral $\int\limits_0^1 \frac{\sin x}{x} dx$. A plot of the integrand (figure 6.4) shows that most of the contribution to the integral comes from the region close to the origin.

The function $\sin(x)/x$ is also known as the sinc function. It plays a role in the Fraunhofer diffraction pattern of a single slit (see example 3.1). Close to the origin, a series expansion of $\sin(x)$ is justified.

Expanding the integrand about the origin, we get:

$$\int\limits_0^1 \frac{\left(x - \frac{x^3}{3!} + \frac{x^5}{5!} - \right)}{x} dx$$

(6.90)

$$= \int\limits_0^1 dx - \int\limits_0^1 \frac{x^2}{6} dx + \int\limits_0^1 \frac{x^4}{120} dx - .. = 1 - \frac{1}{18} + \frac{1}{600} - .. = 0.9461.$$

The advantage of the series expansion procedure is that we can easily judge how many terms need to be included in the series to achieve a certain accuracy. The numerical value obtained in (6.90) is accurate to four decimal places. When we use the quadrature formulae, it is difficult to decide the step size required to achieve a

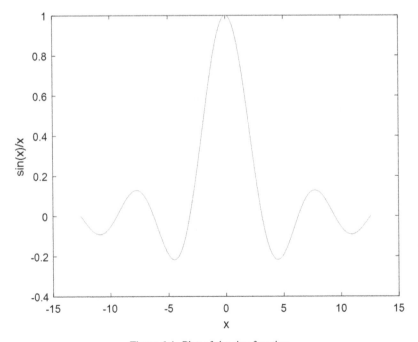

Figure 6.4. Plot of the sinc function.

certain accuracy, since the error depends on some derivative of the function in addition to its dependence on the step size, h.

The series expansion is only useful if we can truncate the series within a few terms and still obtain reasonably accurate results.

Example 6.10

Evaluate $\int\limits_0^{1/2} (1 + x^4)^{1/2}\, dx$ accurate to four decimal places.

Answer: a binomial expansion is justified if x < 1. Here, the domain of integration is from zero to ½, so we are justified in expanding the integrand binomially:

$$\int\limits_0^{1/2} (1 + x^4)^{1/2} = \int\limits_0^{1/2} \left(1 + \frac{x^4}{2} + \frac{1}{2}\left(-\frac{1}{2}\right)\frac{x^8}{2} + \frac{1}{2} \times \frac{1}{2} \times \frac{3}{2} \times \frac{x^{12}}{6} - .. \right) dx.$$

Integrating the terms in the binomial expansion gives us:

$$= \left[x + \frac{x^5}{10} - \frac{x^9}{72} + \frac{x^{13}}{208} - \right]_0^{1/2}.$$

Note that after we substitute the limits, all the terms in the series are increasing powers of ½. Therefore, it is easy to judge the number of terms required to achieve an accuracy of four decimal places ($\sim 1/2^{13}$). We can stop at the fourth term in the series to get the numerical value 0.5031.

6.11 The Fourier transform

Most of this chapter has been about numerical integration techniques. While there are innumerable occasions on which a physicist has to evaluate an integral, we will mention one integral that occurs most frequently in many areas of physics, which is the Fourier transform. The partition function is the other integral that crops up frequently (but only in statistical mechanics).

The one-dimensional Fourier transform is defined by:

$$F(k) = \int_{-\infty}^{\infty} f(x)\exp(-ikx)dx. \tag{6.91}$$

If the function f is a function of time, then x is replaced by t in (6.91) and k is replaced by ω. The Fourier transform appears in many areas of physics. For instance, the intensity of a Fraunhofer diffraction pattern due to an aperture is the square of the Fourier transform of the aperture function[3]. The atomic form factor is the Fourier transform of the electron density within an atom. The reciprocal lattice, which occurs in the study of x-ray diffraction, is the Fourier transform of the direct lattice. The Fourier transform is also important in quantum mechanics.

---

[3] See *Optical Physics* by Ariel Lipson, Stephen Lipson, and Henry Lipson, 4th edn, Cambridge University Press, Cambridge, 2011 for a detailed discussion of Fourier optics.

We will just mention one example of a Fourier transform in which a numerical evaluation of the integral is almost necessary. The Fraunhofer diffraction pattern due to a circular aperture is the two-dimensional Fourier transform of the aperture function. This is the case because we add up the Huygens' spherical wavelets emanating from different points within the aperture.

The two-dimensional Fourier transform is defined by:

$$F(k_x, k_y) = \int_{-\infty}^{\infty} \int_{-\infty}^{\infty} f(x, y)\exp(-i(k_x x + k_y y))dxdy. \tag{6.92}$$

Since the aperture is circular ($f(x,y)$, which is the aperture function, is one within the aperture and zero outside), we transform the integral to polar coordinates, using the relations:

$$x = r \cos \theta, \quad y = r \sin \theta. \tag{6.93}$$

On the screen, the polar coordinates are defined by:

$$k_x = \rho \cos \varphi, \quad k_y = \rho \sin \varphi. \tag{6.94}$$

By substituting (6.94) and (6.93) in (6.92), we obtain:

$$F(\rho, \varphi) = \int_0^{2\pi} \int_0^a \exp(-i\rho r \cos (\theta - \varphi))rdrd\theta. \tag{6.95}$$

Equation (6.95) follows from the fact that the aperture function is only nonzero within the aperture and zero outside. The following program evaluates the integral in (6.95) for $\phi = 0$.

```
'Diffraction pattern of Circular aperture';
clear;
a=1;
r=0:0.01:a;
theta=0:0.01:2*pi;
rho=0:0.01:10;

for k=1:size(rho,2)
    sum(k)=0;
for m=1:size(r,2)
    for n=1:size(theta,2)
      sum(k)=sum(k)+exp(-1i*r(m)*rho(k)*cos(theta(n)))*r(m)*0.01*0.01;
    end
end
end
plot(rho,sum)
```

Note that we have used i in the program to denote the square root of minus one. The default value of i in a MATLAB program is the square root of minus one, but if i is assigned another value by a program, the assigned value will override the default value.

The Fourier transform (and hence the intensity of the diffraction pattern) is found to be zero at $\rho = 3.83$ (figure 6.5).

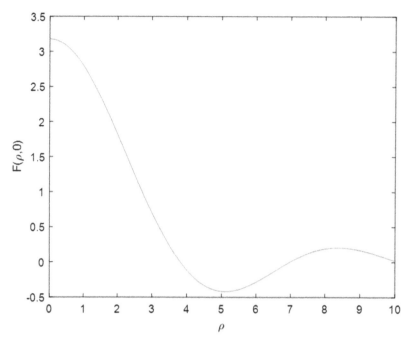

Figure 6.5. The Fourier transform of a circular aperture.

This corresponds to an angle of $\frac{\rho}{k_0} = \frac{3.83\lambda}{2\pi a} = \frac{0.61\lambda}{a}$.

You must have seen this result as part of a discussion of the resolving power of telescopes.

An important fact implicit in (6.91) and (6.92) is that the Fourier transform is, in general, a complex function. However, in many cases, symmetry considerations mean that the Fourier transform is a real function, as seen below.

Example 6.11

Show that for real functions $f(x)$:

(i) the Fourier transform is always real, if $f(x)$ is an even function and purely imaginary, if $f(x)$ is an odd function.

Answer:

(i) From the definition of the Fourier transform:

$$F(u) = \int_{-\infty}^{+\infty} f(x)e^{-ikx}dx = \int_{-\infty}^{+\infty} f(x)\cos(kx)dx - i\int_{-\infty}^{+\infty} f(x)\sin(kx)dx. \quad (6.96)$$

Since the sine function is an odd function, the second integral will vanish if $f(x)$ is even (the product of the two functions will be odd in that case).

(ii) If $f(x)$ is odd, the second integral will be nonzero, but the first integral will vanish, since it involves the product of an odd function $f(x)$ with an even function (the cosine function).

The above result is important in the context of the numerical evaluation of the Fourier transform. In some cases, due to rounding errors, the imaginary part of the Fourier transform does not sum to zero; however, we know from symmetry considerations that the Fourier transform is supposed to be real. In such cases, we can simply ignore the imaginary part of the Fourier transform and take it to be completely real.

The Fourier transform also plays a role in experimental work. In many cases, experimental data may be corrupted by noise. In such cases, we can 'filter' the signal by removing the noise from the Fourier spectrum of the signal. Since experimental data will have a discrete set of points, the discrete Fourier transform is useful in this context.

The discrete Fourier transform (DFT) is defined by:

$$F(\omega) = \frac{1}{M} \sum_{t=0}^{M-1} f(t) \exp\left(-\frac{i\omega t}{M}\right)$$ (6.97)

$$for\ \omega = 0,\ 1,\ 2,\ ......\ M-1.$$

Here, M is the total number of experimental data points. Note that the integral of the continuous Fourier transform is replaced by a summation. The DFT satisfies the properties of the continuous Fourier transform; in addition, it satisfies some additional properties (which are not present in the continuous Fourier transform).

One such property is the periodic nature of the DFT. It can easily be shown that $F(\omega) = F(\omega + M)$.

Proof:

$$F(\omega + M) = \frac{1}{M} \sum_{t=0}^{M-1} f(t) \exp\left(-\frac{i(\omega + M)t}{M}\right)$$

$$= \frac{1}{M} \sum_{t=0}^{M-1} f(t) \exp\left(\frac{-i\omega t}{M}\right) \exp(-iMt)$$

$$= \frac{1}{M} \sum_{t=0}^{M-1} f(t) \exp(-i\omega t) = F(\omega).$$ (6.98)

MATLAB has built-in functions that implementing the discrete Fourier transform. The function 'fft' implements the fast Fourier transform, which is a fast algorithm that evaluates the discrete Fourier transform. The function 'fft2' calculates the two-dimensional fast Fourier transform.

The periodic nature of the DFT helps us to formulate an algorithm known as the fast Fourier transform, which is an efficient and fast method for evaluating the discrete Fourier transform.

Due to the periodic nature of the DFT, the spectrum of an image can appear scrambled. A low-frequency component can masquerade as a high-frequency component.

Note that:

$$F(M-1) = F(-1).$$

$\omega = -1$ is actually a low frequency, but because of the periodic property of the DFT, it appears at $M-1$, which would be the highest possible frequency in an M-point transform.

The MATLAB function 'fftshift' can unscramble the FFT. It shifts the zero frequency component to the centre of the spectrum.

6.12 Numerical integration using MATLAB

MATLAB has a number of built-in functions that can be used for numerical integration. A few of them are tabulated below:

MATLAB function	Technique used
Integral	Global adaptive quadrature (1-D)
integral2	Double integral
integral3	Triple integral
Quadgk	Gauss–Kronrod quadrature
Trapz	Trapezoidal rule
Quad	Simpson's quadrature

Exercises

6.1. (a) Show that the integral $\int_1^\infty \frac{\cos x}{1+x^2} dx$ converges by comparing it with another integral which can be evaluated analytically.
   (b) Use an appropriate variable transformation, i.e. substitution, to make the domain of the integral finite.
   (c) Use the two-point Gaussian quadrature method to evaluate the value of the integral.

6.2. The equation of a certain curve is $r = \sin(2\theta)$, where r and θ are the plane polar coordinates. Compute the total length of the curve using the two-point Gaussian quadrature method.

6.3. Study the program given below:

```
Line number command
1      x0=0;
2      h=0.0001;
3      x1=x0+h;
4      x2=x0+2*h;
5      n=3/2;
6      eta=30.7333333333;
7      g=0;
8      while x0<1000
9      f0=(x0^n)/(1+exp(x0-eta));
10     f1=(x1^n)/(1+exp(x1-eta));
11      f2=(x2^n)/(1+exp(x2-eta));
12      g=g+(h/3)*(f0+4*f1+f2);
13      end
```

(a) What is the program trying to compute?

(b) Which numerical method is being used in the computation?

(c) Point out any errors in the program that give rise to an incorrect output.

(d) Specify the corrections which have to be incorporated in this program to yield the desired output. In addition, specify the line numbers where the corrections have to be incorporated.

6.4. (a) Prove that $\int_0^1 \frac{e^{-x} - 1 + x}{x} dx$ converges.

(b) Evaluate the above integral using the two-point Gaussian quadrature formula

6.5. If $I_1 = \int_0^{\frac{\pi}{4}} \frac{dx}{x^2 + \sin^2 x}$ and $I_2 = \int_0^{\frac{\pi}{4}} \frac{dx}{x^2 + \cos^2 x}$, which integral will have a larger value? You need not evaluate the integral, but give a brief explanation of your answer.

6.6. Even though we mentioned in section 6.4 that the derivation of a five-point or six-point rule would be tedious, a closer look at the augmented matrices used to derive Simpson's rule reveals a pattern that can be exploited to automate the process of deriving quadrature rules with large numbers of points.

Write a MATLAB program to automate the derivation of the augmented matrix for an n-point rule using the following steps as clues:

(i) Note that the first row of the augmented matrix (see (6.30) and (6.39)) has all ones in all the columns, except for the last column for $(n-1)h$. When we automate the procedure for the derivation of quadrature rules, we can ignore the h and just place the coefficient of h in the last column (h can be appended once the program has given the values of each of the weights).

(ii) All entries in the first column are zero except for the element in the first row.

(iii) The second row has entries (starting from the second column) that follow 1, 2, ..$(n-1)$. The entry in the last column of the second row is $(n-1)^2/2$. This entry is obtained by integrating x between the limits zero and $(n-1)h$.

(iv) Similarly, the entries in the third row are 0, 1^2, 2^2, 3^2, ...$(n-1)^2$, $(n-1)^3/3$.

(v) Therefore, in general, the entries in the ith row of the augmented matrix are:

0, 1^{i-1}, 2^{i-1}, 3^{i-1}, ...$(n-1)^{i-1}$, $(n-1)^i/3$.

Once this program has generated the augmented matrix, we can use the program for Gauss elimination (see section 5.1) to obtain all the weights. Test your program by determining the weights for Simpson's rules.

6.7. Test the convergence of $\int_0^\pi \frac{d\theta}{\sqrt{\theta} + \sin \theta}$ by comparing it with a suitable integral (a direct comparison test). Using the result, obtain an upper bound for the value of this integral.

6.8. Evaluate the integral $\int_{-\pi}^{\pi} \frac{\sin x}{x} dx$ using the two-point Gauss–Legendre quadrature rule.

6.9. You have a root-finding program (function), which when called repeatedly (i.e. recursively) will determine the roots of a polynomial. If a different root is determined each time the function is called, what is the minimum number of times the program has to be called to determine all the roots of a Legendre seventh-degree polynomial? Note that you can minimize the number of function calls by using the properties of Legendre polynomials.

6.10. What is the numerical value obtained if you evaluate $\int_{-1}^{1} (5x^3 - 3x) dx$ using the two-point Gaussian quadrature formula? Is this value exact, or is it approximate? Explain your answer. Hint: you do not have to actually perform any calculations to answer this question!

6.11.
(a) Prove that $\int_{0}^{2} \frac{dx}{(1+x)\sqrt{x}}$ converges.

(b) Evaluate the above integral using the two-point Gaussian quadrature formula.

Chapter 7

Monte Carlo integration

7.1 Error in multidimensional integration

The quadrature methods discussed in chapter 6 can be extended in a straightforward manner to two or more dimensions. The integral $\int_a^b \int_c^d f(x, y)dxdy$ represents the volume under the surface $f(x,y)$.

Instead of elaborating on the implementation of quadrature methods in higher dimensions, we instead focus on how the error of the quadrature methods depends on the dimension. Using the rectangular approximation, we will examine how the error depends on the dimension.

The error in a double integral evaluated using the rectangular approximation is given by:

$$\Delta_i = \int_{x_i}^{x_{i+1}} \int_{y_i}^{y_{i+1}} f(x, y)dxdy - f(x_i, y_i)\Delta x \Delta y. \tag{7.1}$$

The second term in the right-hand side of (7.1) is the value of the integral, while the first term is the value of the integral over a small rectangle with sides Δx and Δy. To evaluate the error in the rectangular approximation, we use the Taylor expansion of $f(x,y)$ about (x_i, y_i) and keep only the terms up to the first derivative, as the leading term in the error is proportional to the first derivative (the function is treated as a constant over the small interval in the rectangular approximation).

Using the Taylor expansion of $f(x,y)$ and implementing the rectangular approximation in (7.1), we obtain:

$$\Delta_i = \int_{x_i}^{x_{i+1}} \int_{y_i}^{y_{i+1}} (x - x_i)\frac{\partial f}{\partial x}\bigg|_{x_i} dxdy + \int_{x_i}^{x_{i+1}} \int_{y_i}^{y_{i+1}} (y - y)\frac{\partial f}{\partial y}\bigg|_{y_i} dxdy \tag{7.2}$$

doi:10.1088/978-0-7503-3791-5ch7

$$\frac{\Delta x^2}{2} \frac{\partial f}{\partial x}\bigg|_{x_i} \Delta y + \frac{\Delta y^2}{2} \frac{\partial f}{\partial y}\bigg|_{y_i} \Delta x. \tag{7.3}$$

We wish to only perform n function evaluations to evaluate the integral over the entire area of integration, A; then

$$\frac{A}{n} = \Delta x \Delta y. \tag{7.4}$$

By assuming $\Delta x = \Delta y$, we find the total error (i.e. over the entire range of integration), which is:

$$\text{Total error} = n\Delta_i \propto n\Delta x^3 = n\left(\frac{A}{n}\right)^{3/2} = n^{-1/2} \tag{7.5}$$

According to (6.20), the error in the rectangular approximation over one step of width h is proportional to h^2. The total error over the entire range of integration will be nh^2. From (6.18), we have: $h = \frac{(b-a)}{n}$; therefore, the total error in one dimension is:

$$\propto nh^2 = \frac{(b-a)^2}{n}. \tag{7.6}$$

Therefore, in one dimension, the error is proportional to n^{-1}, while in two dimensions, we see from (7.5) that it is proportional to $n^{-1/2}$. Therefore, it is reasonable to suppose that in d dimensions, the error in the rectangular approximation will be proportional to $n^{-1/d}$. Consequently, the error increases with increasing dimensions. A similar conclusion holds for the trapezoidal and Simpson rules.

We encounter higher-dimensional integrals in statistical mechanics when we evaluate the partition function. For N particles moving in three-dimensional space, the partition function is an integral over the $6N$ dimensions of phase space. If we try to evaluate such integrals by quadrature methods, the error is likely to be very high. We can reduce the error by increasing the value of n, but that would increase the number of function evaluations and slow down the program.

We need an alternate approach that keeps the error within reasonable bounds for the same number of function evaluations. A class of methods known as Monte Carlo methods does just that.

7.2 Monte Carlo integration

Monte Carlo is part of the principality of Monaco, which is situated on the northern coast of the Mediterranean sea. Monte Carlo is famous for its casinos, which are synonymous with gambling. The subject of probability has a close connection with gambling. Monte Carlo integration is a probabilistic method that evaluates the area under a curve. The technique can easily be extended to higher dimensions. Probability is necessary to understand the behaviour of many physical systems.

In classical physics, probability enters the discussion because we may not have complete information about the system. However, nature seems to be intrinsically probabilistic at the quantum level. We will discuss probability and its uses in physics further in chapter 8. For the present, we confine ourselves to a discussion of how it can be used to evaluate an integral, i.e. the area under a curve.

The fundamentals of Monte Carlo integration can be understood from a simple thought experiment. Imagine that you are standing at the edge (point P in figure 7.1) of a rectangular field that is fenced off. You are asked to find the surface area of the water in a pond (shaded blue in figure 7.1) that is situated within the field. While the field has a rectangular shape, the boundary of the pond is irregular. You can adopt a Monte Carlo, i.e. probabilistic method to determine the area. Imagine you start throwing stones with random velocities and at random angles to the horizontal, while ensuring that the stones always land within the field. Some stones will land inside the pond, while others will miss the pond but land within the field.

If you throw a total of N stones within the field and n of them land within the pond, then clearly:

$$\frac{\text{Area of pond}}{\text{Area of field}} = \frac{n}{N} \tag{7.7}$$

From (7.7), we have:

$$\text{Area of pond} = \frac{n}{N}\text{Area of field} \tag{7.8}$$

Since the area of the field is known, as it is a rectangle, the area of the pond can be calculated from (7.8).

A similar approach can be used to calculate the area under a curve.

Let us say we have to evaluate the integral $I = \int_a^b f(x)dx$. We draw a rectangle such that one side lies along the x-axis with a length of $(b-a)$ and its breadth is such

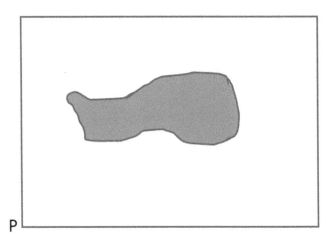

Figure 7.1. Using the Monte Carlo method to find the area of a pond.

that the curve $f(x)$ lies below the top of the rectangle from $x = a$ to $x = b$. In other words, the rectangle, also known as the bounding rectangle, completely encloses the curve from $x = a$ to $x = b$.

To evaluate the area under the curve in figure 7.2, we can follow the procedure we followed to calculate the area of the pond. However, instead of throwing stones, we generate N random points within the bounding rectangle shown in figure 7.2. Among the N points, let n be the number of points below the curve $f(x)$. Then, following (7.8), the area under the curve is given by:

$$I = \int_a^b f(x)dx = \frac{n}{N} \times \text{Area of rectangle} \qquad (7.9)$$

The error in this method can be expected to depend upon the value of N. We expect the error to reduce as N increases. It should also depend upon $f(x)$. We will make a precise calculation of the error, but first, we will discuss a few details of the implementation of (7.9).

To generate N random points, we need to randomly generate the x and y coordinates of these points such that they always lie within the rectangle. The MATLAB function RAND can be called repeatedly (i.e. N times) to generate the x and y coordinates. If the function RAND is used without any input arguments, it returns a single number in the range from zero to one. If h is the height of the bounding rectangle, the following commands can be used within a FOR loop to generate N points within the rectangle:

$x = a + (b - a)*\text{rand};$
$y = h*\text{rand};$

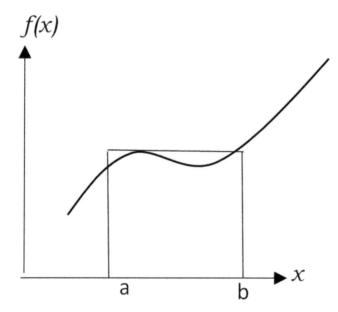

Figure 7.2. Area under a curve using Monte Carlo integration.

We will provide the complete program for implementing Monte Carlo integration, but first let us try another version of the Monte Carlo method. In this version, known as the sample mean method, (7.9) gets modified to:

$$I = \int_a^b f(x)dx = <f(x)> (b - a). \qquad (7.10)$$

In the sample mean method, the area under the curve is equal to the average value of the function $f(x)$ over the domain (a,b) multiplied by the width of the domain, i.e. $(b-a)$. The average value of the function is calculated using its value at N points, which are randomly selected from within the interval (a,b):

$$<f(x)> = \frac{1}{N}\sum_{i=1}^{N} f(x_i). \qquad (7.11)$$

The program given below evaluates the integral $\int_0^1 \sqrt{1 - x^2}\,dx$, which has an exact value of $\pi/4$.

```
'program to do Monte Carlo integration';
'hit or miss method';
count=0;
n=1000;
'n is the number of trials';
sum=0;
for i=1:n
    x=rand;
y=rand;
f=(1-x^2)^0.5;
sum=sum+f;
if y<=f
    count=count+1;
end
end
mc=count/n;
sm=sum/n;
```

The variable *mc* will contain the value of the integral obtained by the bounding rectangle method, while *sm* will have the value obtained by the sample mean method. On running the program once, I obtained the value 0.7900 using the bounding rectangle method. Since the points are randomly chosen, the number of points below the curve will vary each time the program is run, leading to different values for the integral.

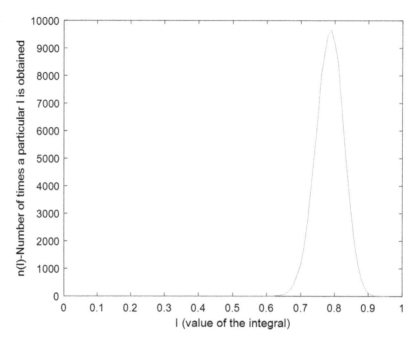

Figure 7.3. Distribution of values obtained for the integral discussed above.

To evaluate the integral more accurately, we run the program many times and plot the distribution of values obtained (figure 7.3)

We see that we get values of the integral that are normally distributed around the exact value. This is one way of obtaining the exact value of the integral using Monte Carlo methods. To obtain the distribution shown in figure 7.3, we set N (the number of random points) to 100 and repeated the evaluation of the integral 100 000 times.

Figure 7.3 gives us a clue about the error involved in Monte Carlo integration. We will use a rigorous procedure to estimate the error introduced by Monte Carlo integration.

7.3 Error estimate for Monte Carlo integration

Let us assume that we use n random points to estimate the integral of a function by the Monte Carlo method. To get an accurate value, we repeat the procedure m times. This means that we are sampling the function at mn points. We will use the sample mean method to estimate the error in our procedure. Let M_α be the mean value of the function in the αth determination of the integral.

Since we are using the sample mean method, we have:

$$M_\alpha = \frac{1}{n} \sum_{i=1}^{n} f_{\alpha i}. \tag{7.12}$$

Let $\bar{M}$ be the mean value of the function obtained over mn points; then

$$\bar{M} = \frac{1}{mn} \sum_{\alpha=1}^{m} \sum_{i=1}^{n} f_{\alpha i}. \tag{7.13}$$

The variance of the values obtained over the m trials will be:

$$\sigma_m^2 = \frac{1}{m} \sum_{\alpha=1}^{m} (M_\alpha - \bar{M})^2 = \frac{1}{m} \sum_{\alpha=1}^{m} \left(\frac{1}{n} \sum_{i=1}^{n} f_{\alpha i} - \bar{M} \right) \left(\frac{1}{n} \sum_{j=1}^{n} f_{\alpha j} - \bar{M} \right). \tag{7.14}$$

The product on the right-hand side of (7.14) contains two kinds of term: those for which $i=j$ and those for which $i \neq j$. The indices i and j correspond to two different trials. Since the points are chosen randomly and independently in each trial, the cross-terms (those for which $i \neq j$) are equally likely to be positive or negative and cancel out, reducing the double summation over i and j to a single summation over i. Therefore, we have:

$$\sigma_m^2 = \frac{\sigma^2}{n}. \tag{7.15}$$

Here, σ^2 is the variance of the function, i.e. the mean square deviation of the function from its mean over mn samples. Equation (7.15) gives us an estimate for the error in the Monte Carlo method. We see that to reduce the error, we need to either increase n (as been mentioned previously) or reduce σ^2. Reducing σ^2 may appear impossible, since it is the variance of the integrand. However, this is precisely what the importance sampling Monte Carlo method does.

The fact that the error in the Monte Carlo method depends on the variance of the function is intuitively obvious. Suppose that we are integrating $\int_0^2 x^4 dx$. The integral has a larger contribution in the interval (1,2) compared to its contribution in the interval (0,1). However, the normal Monte Carlo method, which uses a uniform distribution, gives equal importance (i.e. roughly the same number of points) to both regions. In this case, the integrand has a large variation over the region of integration; therefore, the error will be greater. Choosing a $p(x)$ that gives more weight to the interval (1,2) will work better for this integral.

A remarkable fact about (7.15) is that its error is independent of the dimension. As a result, for higher dimensions, the Monte Carlo method is expected to have an error that is smaller than the error introduced by the quadrature methods discussed in chapter 6. You may recall from section 7.1 that the error increases with dimension for the quadrature methods.

Figure 7.4 shows the errors introduced by determining the integral $\int_0^1 4\sqrt{1 - x^2}\, dx$ using the sample mean and hit-or-miss (bounding rectangle) methods. Equation (7.15) was derived for the sample mean method. Furthermore, it should be noted from figure 7.4 that the error introduced by the sample mean method is less than that introduced by the hit-or-miss method. This is true in general.

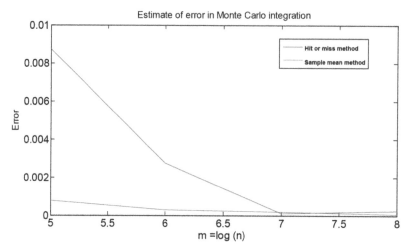

Figure 7.4. Errors introduced by Monte Carlo methods as a function of *n*.

7.4 Importance sampling Monte Carlo

As can be seen from (7.15), the error in the Monte Carlo method can be reduced if we reduce the variance of the integrand. But how do we do this? To reduce the variance, rewrite the original integral as:

$$I = \int_a^b f(x)dx = \int_a^b \frac{f(x)}{p(x)} p(x)dx. \tag{7.16}$$

In (7.16), we multiplied and divided the integrand by $p(x)$. In the importance sampling method, we treat the integrand as $f(x)/p(x)$, while $p(x)dx$ is considered to be the probability distribution for the values of x chosen to evaluate the integral. If we choose $p(x)$ to resemble $f(x)$, then the integrand $f(x)/p(x)$ will have a lower variance, as it will become almost constant over the interval (a,b). However, $p(x)$ has to be normalised over the interval (a,b), which means it should be integrable.

How can we generate many values of x that follow a certain distribution $p(x)$? We can exploit the fact that we already have a random number generator that can generate numbers with a uniform distribution in the interval $(0,1)$. We let r denote a random number in the interval $(0,1)$. Since $p(r)$ is a uniform distribution, we have $p(r) = 1$ for all values of r in the interval $(0,1)$ and $p(r) = 0$ outside this interval.

Assuming that the mapping from r to x (which we need) is a one-to-one mapping, we have:

$$p(r)dr = p(x)dx. \tag{7.17}$$

If we multiply both sides of (7.17) by n (the total number of points), the resulting equation can be interpreted as 'all values of r in the range r to $r+dr$ are mapped to values of x in the range x to $x+dx$'.

Using the fact that $p(r) = 1$ and integrating both sides of (7.17), we have:

$$r = \int_{-\infty}^{x} p(x')dx'. \tag{7.18}$$

If $p(x)$ is non-zero only in the same range as r, then the lower limit in (7.18) can be changed to zero. Equation (7.18) is not useful as it stands; we need to invert (7.18) to get the x value corresponding to a given r. Therefore, we invert (7.18) after evaluating the integral over x'. Hence, this method is also known as the 'inverse transform' method.

We now present an example that illustrates the importance sampling method (inverse transform).

Example 7.1

Evaluate the integral $\int_0^1 \exp(-x^2)dx$ using the importance sampling method.

Answer: In the importance sampling technique, we need to choose a $p(x)$ that resembles the integrand—at least over the integration domain. Let us try $\exp(-x)$, which has the same behaviour as the integrand over the small interval $(0,1)$.

First, we need to normalise $\exp(-x)$, so we multiply it by a normalisation constant A, such that:

$$\int_0^1 A \exp(-x)dx = 1. \tag{7.19}$$

We obtain the normalisation constant from (7.19):

$$A = \frac{1}{1 - e^{-1}}. \tag{7.20}$$

We now apply the inverse transform method (7.18) for this function:

$$r = \int_0^x A \exp(-x')dx' = A(1 - \exp(-x)). \tag{7.21}$$

By substituting for A in (7.21) and inverting equation (7.21), we get:

$$1 - r(1 - e^{-1}) = \exp(-x) \tag{7.22}$$

or

$$x = \ln\left[\frac{1}{1 - r(1 - e^{-1})}\right]. \tag{7.23}$$

Equation (7.23) is now in a readily implementable form. To generate n values of x, which follow the exponential distribution, we generate n values of r (using the random number generator) and then substitute those values of r into (7.23) to get the corresponding values of x. Verify that $r = 0$ maps to $x = 0$ and $r = 1$ maps to $x = 1$ according to (7.23). The values of r in the interval $(0,1)$ will be mapped to values of x in the same interval, but more values of x will be close to zero than those close to one, as $p(x) = A \exp(-x)$.

7.5 The Box–Muller method

We now describe the Box–Muller method, which is useful for generating a Gaussian/normal distribution. Suppose we would like to generate the normal distribution:

$$p(x)dx = \frac{1}{\sqrt{2\pi\sigma^2}}\exp\left(-\frac{x^2}{2\sigma^2}\right)dx. \tag{7.24}$$

Instead of generating this distribution, we can generate the two-dimensional[1] distribution:

$$p(x, y)dxdy = \frac{1}{2\pi\sigma^2}\exp\left(-\frac{(x^2 + y^2)}{2\sigma^2}\right)dxdy. \tag{7.25}$$

The obvious next step is to transform this to polar coordinates: and

$$r = \sqrt{x^2 + y^2} \text{ and } \theta = \tan^{-1}\left(\frac{y}{x}\right) \tag{7.26}$$

The two-dimensional distribution in (7.25) is now transformed to:

$$p(r, \theta)rdrd\theta = \frac{1}{2\pi\sigma^2}\exp\left(-\frac{r^2}{2\sigma^2}\right)rdrd\theta. \tag{7.27}$$

Suppose we want to generate a distribution with a variance equal to one and define:

$$\frac{r^2}{2} = \rho. \tag{7.28}$$

Equation (7.27) can now be written as:

$$p(\rho, \theta)d\rho d\theta = \frac{1}{2\pi}\exp(-\rho)d\rho d\theta. \tag{7.29}$$

We can generate values of p that follow an exponential distribution (following the procedure in example 7.1) and generate values of θ with a uniform distribution from zero to 2π.

The values of p and θ we have can be transformed back to (x,y) using the equations:

$$x = r\cos\theta \quad y = r\sin\theta \tag{7.30}$$

or

$$x = \sqrt{2\rho}\cos\theta \quad y = \sqrt{2\rho}\sin\theta. \tag{7.31}$$

The values of x and y that we obtain by this process will individually satisfy Gaussian distributions.

---

[1] A similar trick is adopted to integrate the Gaussian function over the limits $-\infty$ to $+\infty$.

7.6 The Metropolis algorithm

We now discuss two other methods for generating probability distributions. Even though we came across the problem of generating probability distributions in the context of the importance sampling method, probability distributions have many applications in physics (especially in statistical mechanics). For instance, we might be simulating the behaviour of a collection of particles and might want their initial velocities to have a certain distribution.

A very simple (but crude) method of generating a set of values having a certain distribution is to use the hit-or-miss (also called the acceptance-rejection) method. This is the method discussed earlier in section 7.2 in the context of Monte Carlo integration. Suppose that we wish to generate values of x in the interval (a,b), which follow the distribution $p(x)$; we draw a bounding rectangle around $p(x)$ (see figure 7.2). We then generate, say, n points (x,y) using a random number generator. For each of these random points, we check whether $y < p(x)$. If this condition is satisfied, we accept that value of x and add it to an array. If the condition is not satisfied, we reject that value (i.e. we do not store it in the array). If we require n values of x in the array, we need to generate random points until the required number is reached.

Figure 7.5 shows a million values of x that follow an exponential distribution, generated using the hit-or-miss method. The plot shows the number of values of x, i.e. $n(x)$, that lie in the interval x to $x + dx$. To verify the exponential nature of the distribution,

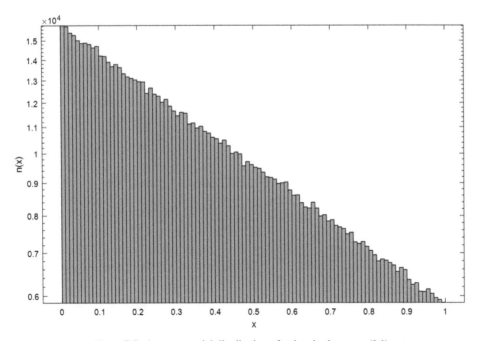

Figure 7.5. An exponential distribution of points in the range (0,1)

the scale along the y-axis has been made logarithmic; therefore, the plot should be a straight line with a negative slope. Figure 7.5 shows that this is indeed true.

We now describe a more sophisticated way of generating a distribution, known as the Metropolis algorithm. Contrary to what you might imagine, 'Metropolis' in this context refers not to a city but the first author[2] of the original paper that described this algorithm!

This algorithm works as follows:

   i. Choose an initial value of x, say x_0. If we are generating, say, a Gaussian distribution, x_0 can be the peak of the Gaussian.

  ii. Choose a trial value of x, say x_{trial}. As in the accept–reject or hit-or-miss method, we need to decide whether to accept this trial value of x. Normally, $x_{trial} = x_0 + \delta$, where δ is a random number in the interval $(-\delta_{max}, +\delta_{max})$.

  iii. Calculate the ratio of the probabilities $w = \frac{p(x_{trial})}{p(x_0)}$.

  iv. If the ratio $w > 1$, accept the trial value and append it to an array that will contain all the accepted values.

  v. If $w < 1$, generate a random number r in the range zero to one.

  vi. If $w >= r$, accept x_{trial} and append it to the array.

  vii. If $w < r$, reject the trial value and append the existing value of x (i.e. x_0), to the array. Note that rejections lead to repetitions in the values of x. This is fine, as those values of x are more probable than nearby values.

  viii. Repeat steps (ii) to (vii) until a sufficient number of values of x are available. In subsequent iterations, the previous value of x replaces x_0 in step (ii).

How do we know how many values of x are sufficient to obtain the required distribution? A rule of thumb is to choose the maximum step size (δ_{max}) such that one third to half of the trial values are accepted. An alternative criterion is to monitor the standard deviation of the values of x in the array and stop the process if the standard deviation is close to the required value.

Figure 7.6 shows a normal distribution with a mean of zero and a standard deviation equal to one generated using the Metropolis algorithm.

7.7 Random number generators

We have discussed Monte Carlo methods in some detail. Their importance to applications in science and engineering should not be underestimated. We will devote the next chapter to discussing these applications. However, we also need to look closely at how computers generate random numbers.

A moment's reflection will tell you that a computer cannot generate a truly random number[3]. This is because a digital computer is a deterministic machine that

---

[2] In total, this paper had five authors. The original paper that described this algorithm, published in 1953, can be accessed at: https://bayes.wustl.edu/Manual/EquationOfState.pdf.

[3] The famous mathematician John Von Neumann said that 'Anyone who considers arithmetical methods of producing **random** digits is, of course, in a **state of sin**'.

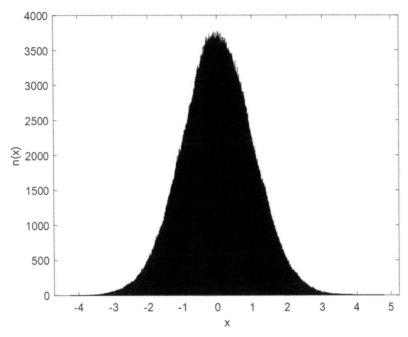

Figure 7.6. A normal distribution generated using the Metropolis algorithm.

applies deterministic rules. How can it then generate a truly random number? The 'random' numbers spewed out by a computer are pseudo-random numbers. A good pseudo-random number generator will generate a series of numbers that does not have any obvious pattern and appears random at first glance. This is especially true if we do not know the algorithm used to generate these numbers.

While Von Neumann was critical of arithmetical methods for generating random numbers (see footnote 3), he suggested one such method, known as the middle-square method. In the middle-square method, we choose a four-digit number. This number is known as the 'seed' of the random number generator. All computer-based methods for generating random numbers require a seed. Square the four-digit number and choose the middle four digits. Add leading zeros if the resulting number has less than eight digits. Now choose the middle four digits to obtain the next random number in the sequence.

As an illustration, let us say we start with the four-digit number 2345. On squaring 2345, we get 5499025. Since this is a seven-digit number, we add a leading digit to make it 05499025 and choose the four digits in the middle, i.e. 4990; we square this to obtain 24900100. Repeating the process on this number, we get 9001; squaring it gives 81018001. Now we are left with 0180 as the middle four digits. If we collect all the four-digit numbers, we have the sequence: 2345. 4990, 9001, 180, … This is a reasonable sequence, but sometimes the numbers start repeating, or, what is worse, we can sometimes end up with zero.

7.8 The linear congruential method

In this method, the next number in the sequence is generated using the previous number according to the relation:

$$x_{n+1} = (ax_n + c)\mod m, \tag{7.32}$$

where a, c, and m are non-negative integers, and the mod operator gives the remainder after division by m. It is clear that in this method, the longest possible period is m. To come close to the longest period, a and m have to be co-prime. Furthermore, it is better to choose m itself such that it is a prime number. Even if a and m are co-prime, this by itself does not guarantee a long period. For example, if we choose $a = 3$, $c = 5$, and $m = 73$, we get the following sequence of numbers:

$$46, 70, 69, 66, 57, 30, 22, 71, 72, 2, 11, 38, 46$$

which gives us a period of 12 (the same number is repeated after 12 iterations).

To obtain a much longer period, a should be a primitive of the field $GF(m)$[4]. For example, two is a primitive element of GF(13). By choosing $a = 2$, $c = 5$, and $m = 13$, we get:

$$1\ 7\ 6\ 4\ 0\ 5\ 2\ 9\ 10\ 12\ 3\ 11\ 1.$$

This is close to the maximum possible period, as all the elements of GF(13) have been generated except for eight.

However, if we choose, $a = 5$, $c = 0$, and $m = 32$, we get: 26 2 10 18 26, which have a much shorter period of four.

To get a long period, m should be a really large prime number. A detailed discussion of prime numbers would be out of place here. Please refer to books on number theory (see footnote below). Even though a formula for generating a prime number does not exist, efficient methods exist. One way is to generate a number that is likely to be prime and then verify its primality. One such number is a Mersenne prime. Mersenne numbers have the form $2^n - 1$, where n is a natural number. If n is a prime number, then the corresponding Mersenne number is likely to be prime. It should be noted that not all prime values of n lead to Mersenne primes. Mersenne numbers have to be tested for primality.

It should be noted that the period is independent of the first number in the series (also known as the seed). The seed determines all the subsequent numbers in the series, since we are using a completely deterministic algorithm. We would like a different series of numbers to be generated each time the program runs. One way to generate a new seed for each run of the program is to access the system clock. The MATLAB function clock returns six numbers, the first of which is the year, and the last of which is the number of seconds accurate to one millisecond. The time interval

---

[4] A detailed discussion of this point is beyond the scope of this book. The reader is referred to a book on number theory or Galois fields. See for instance, *Elementary Number Theory* by David Burton, 7th edn, McGraw Hill, New York, 2017.

between two runs of the program is likely to be random, so the last number (i.e. the seconds) can be used as the seed.

7.9 Generalised feedback shift register

Since computers operate using bits, a method that uses the binary system suggests itself. One such method is the generalised feedback shift register, which is defined by (7.33):

$$x_n = x_{n-p} \oplus x_{n-q}. \tag{7.33}$$

Here, the symbol $\oplus$ stands for an exclusive OR (XOR) performed bit-wise on the number x written in binary form. If $p > q$, then we need a set of p numbers as the seed. One way to obtain this would be to generate a series of p numbers using another random generator, such as the linear congruential method.

The 'randomness' of a random number generator can be checked by subjecting it to a series of tests. One test could be to plot the distribution of numbers generated by a random number generator. The distribution should be uniform, but a perfectly uniform distribution would also be very suspicious! Furthermore, there should be no apparent pattern. This can be checked by calculating the correlation coefficient defined by:

$$C(k) = \frac{\langle x_{i+k} x_i \rangle - \langle x_i \rangle}{\langle x_i x_i \rangle - \langle x_i \rangle \langle x_i \rangle}. \tag{7.34}$$

Exercises

7.1 Evaluate the integral $\int_0^2 x^2 dx$ using the 'hit-or-miss' Monte Carlo method.
   (a) What is the area of the smallest bounding rectangle which can be used for this integration?
   (b) What is the minimum number of points required for the evaluation of this integral if you want the answer to be accurate to 0.0001?

7.2 Consider the equation given below that generates a sequence of pseudo-random numbers:

$$x_{n+1} = x_n^a \bmod b.$$

Here, a and b are integers, and this equation is iterated to yield a sequence of numbers wherein the $n+1$th number is generated from the nth number.
   (a) What property should be satisfied by b so that this random number generator has a long period?
   (b) Would interchanging the positions of x_n and a improve the performance of this generator (in terms of its period)? Explain your answer. Interchanging the positions means that you now iterate the equation $x_{n+1} = a^{x_n} \bmod b$.

7.3 A student wants to generate a set of values of x which satisfy the distribution

$p(x) = A \exp(-x^2)$, where A is the normalisation constant. To generate this distribution, he substitutes a given value of x into the above expression and determines the frequency of occurrence of that value of x. What is/are the shortcomings and errors in this procedure?

7.4 The integral $\int_0^\pi \frac{dx}{x^2 + \cos^2 x}$ is to be evaluated using the importance sampling Monte Carlo method. Suggest a form of non-uniform probability distribution which you would use to evaluate this integral by the importance sampling Monte Carlo method.

7.5 Consider the following procedure that generates a sequence of two-digit random numbers[5]:

take a two-digit number (whose digits are different). Arrange the digits in descending order and then in ascending order to get two different two-digit numbers, N_1 and N_2, respectively. Subtract the smaller number (N_2) from the larger number (N_1) to obtain a number N_3. Repeat this process iteratively to generate a sequence of numbers (N_3).

What is (are) the drawbacks of this procedure? Would the sequence of numbers (N_3) obtained be random?

7.6 Write a program to generate the coordinates of 100 random points, all of which lie within the triangle whose vertices are at (1,0), (−1,0), and (0,1). Assume that the program can call a function 'rand' to generate a random number in the range zero to one.

7.7 Write a program using MATLAB or pseudo-code to evaluate the area between the curves $y = \sqrt{x}$ and $y = x/3$. Before writing the program, you might want to plot the given curves. In writing the program, assume that you can call a random number generator 'rand' to generate a random number in the range zero to one.

7.8 Suppose you want to evaluate the integral $\int_0^\pi \exp(-x)\sin(x)dx$ by the hit-or-miss Monte Carlo method. Calculate the area of the minimum bounding rectangle which is required to evaluate the integral.

7.9 Assume that you have to generate values of x (in the range −1 to 1) which follow the probability distribution $p(x) = \frac{A}{1+x^2}$ using the inverse transform method.

(a) Determine the constant A.
(b) Obtain the relation between x and r, where r is a random number in the range zero to one.

7.10 You wish to generate a set of values of x which follow the distribution given below:

$p(x) = A(1-x)$ if $0 \leqslant x \leqslant 1$;
$p(x) = 0$ otherwise using

[5] The corresponding procedure for four-digit numbers is known *as Karpekar's routine*. The number 6 174 is a fixed point of this procedure. All numbers eventually converge to this fixed point.

the inverse transform method.

(a) What value should A take?

(b) Derive the relation between x and r, where r is a random number in the range zero to one.

7.11 Write a program in MATLAB or pseudo-code to evaluate the area of overlap between the curves $9x^2 + 25y^2 = 225$ and $x^2 + y^2 - 10x + 16 = 0$ using the 'hit-or-miss' Monte Carlo method. Before writing the program, give a brief explanation of the procedure you will adopt (with a figure if required). In writing the program, assume that you can call a random number generator 'rand' to generate a random number in the range zero to one.

7.12 Write a program in MATLAB or pseudo-code which can be used to generate 100 random points within a circle whose radius is five units and whose centre is at (1,1). The program can call a random number generator which generates random numbers in the range zero to one.

7.13

(a) Name one advantage of the linear congruential method over the generalized feedback shift register method for generating random numbers.

(b) How is the seed for a random number generator normally generated?

IOP Publishing

Computational Methods Using MATLAB®
An introduction for physicists
P K Thiruvikraman

Chapter 8

Applications of Monte Carlo methods

Monte Carlo methods can be used in many areas of science and engineering; however, we will discuss only a few applications in detail, namely: random walks, the Ising model, simulated annealing, and percolation theory.

8.1 Random walks

While the idea of a random walk may appear to be mathematical, it appears in many physical systems (the diffusion of atoms, polymers, etc.). We assume that the reader has heard of random walks. A drunken man (or sailor) is the usual scapegoat employed to explain the basic concept of a random walk. A drunken man starts from a particular point (let's designate this point as the origin). To simplify the discussion, let us discretise time and assume that he takes a step (always of the same length l) to the left or the right with equal probability. Since the steps are taken randomly, the man's position after taking N steps will vary each time the experiment (i.e. random walk) is repeated. In such a situation, we can only ask for the probability $p(x)$ or $p(m)$, that he will be at a particular x after N steps. Let $x = ml$. Let n_1 be the number of steps taken to the right (taken to be the positive x-direction), and let n_2 be the number of steps taken to the left (taken to be negative, as per the usual convention). We then have:

$$N = n1 + n2 \quad \text{and} \quad m = n1 - n2. \tag{8.1}$$

From (8.1), we have:

$$n1 = (N + m)/2 \quad \text{and} \quad n2 = (N - m)/2. \tag{8.2}$$

Since, for a one-dimensional random walk, only two events (a step to the right or a step to the left) are possible, $p(m)$ follows the binomial distribution given by:

doi:10.1088/978-0-7503-3791-5ch8

$$p_N(m) = \frac{N!}{\left[\frac{(N+m)}{2}\right]!\left[\frac{(N-m)}{2}\right]!}.$$

(8.3)

For large values of N, the binomial distribution changes to a normal distribution. The simple random walk described above can easily be analysed, but many questions about random walks can only be answered using numerical simulations. A random number generator is the crucial ingredient required to simulate a random walk.

A self-avoiding random walk is a case of physical interest, as it simulates the configuration of a polymer. In a self-avoiding random walk in two or three dimensions, the walker does not return to a lattice site that has been visited before. When a program performs a self-avoiding random walk, it takes a trial step in one of the four possible directions (if it is a 2D walk). The trial step is accepted only if the site has not been visited before. Very soon, the random walker may end up in an impossible situation, i.e. the random walker has already visited all the nearest neighbours. At this stage, the walk is terminated. It is interesting to repeat the self-avoiding random walk many times and determine the distribution of the walk-lengths obtained. The distribution obtained can then be compared with those obtained experimentally for actual polymers.

To illustrate some important concepts using a random walk, let us start with a simpler system known as the Ehrenfest model (which is not as well known as Schrödinger's cat!). Ehrenfest proposed a simple thought experiment to understand the origin of the second law of thermodynamics. Imagine that there are two dogs. Initially, there are N fleas on one of the dogs, while the other dog is free of fleas. When the simulation starts, one of the fleas is chosen randomly and shifted to the other dog. The two dogs can also be imagined as the two halves of a container separated by a partition. Initially, all the gas molecules or atoms in the container are in one half, while the other half is completely empty. At each time step, a molecule is chosen at random and shifted to the other half. We expect that after a long time, equilibrium will be reached, and each half will have roughly an equal number of molecules (or each dog will have $N/2$ fleas). While the equilibrium state is obvious, it is interesting to use this simulation to study the approach to equilibrium and the fluctuations which are present in the system, even after the equilibrium state is reached. It is also interesting to vary the total number of fleas, N, and study the effect this has on the standard deviation of the final equilibrium state.

Figure 8.1 shows the results obtained for a system with 500 molecules.

Note the rapid decrease in the number from 500 to the final equilibrium value of 250 and the fluctuation about the equilibrium value. To quantify the fluctuation, we can calculate the standard deviation of the number of atoms about the equilibrium value. To calculate the standard deviation, we need to ignore the initial 1 000 values (in figure 8.1, we can ignore the first 1 000 values), as the system has not equilibrated until then. The standard deviation is given as a function of N in table 8.1.

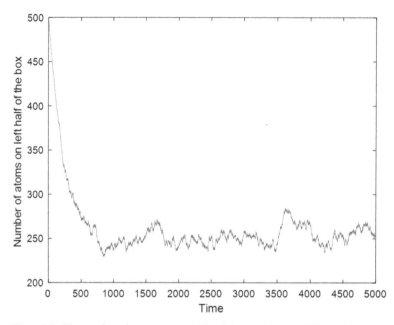

Figure 8.1. The number of atoms on one side of the container at different time steps.

Table 8.1. Standard deviation as a function of N.

Number of particles (N)	Standard deviation	Std dev/N
20	1.7590	0.0879
50	3.4575	0.0691
100	4.1458	0.0414
200	6.3176	0.0315
500	11.936	0.0238
1 000	12.3055	0.0123
2 000	15.1603	0.0076

We notice from table 8.1 that the standard deviation increases with N, but the ratio of standard deviation to N decreases. Theoretically, the ratio of standard deviation to N is expected to be proportional to $1/\sqrt{N}$.

If we fit a straight line to a plot of log(std/N) versus log(N), we get a slope of -0.52, which is close to the theoretically expected value of -0.5. Some error is introduced into the values of standard deviation tabulated in table 8.1, as we need to ignore the initial values wherein the system has not attained equilibrium. Since we have to choose the cutoff point arbitrarily, some deviation from the theoretical result is to be expected. The agreement with the theoretical value is also expected to improve if we run the simulation for larger values of N and for longer time (once equilibrium has been reached). Table 8.1 shows that the fluctuations (std deviation/N) decrease as the system size increases. However, limits on computational facilities may mean that we cannot make N arbitrarily large.

8.2 The Ising model

The Ising model is one of the simplest models that can be used to study phase transitions in magnetic systems. It has been used to study phase transitions in certain non-magnetic systems as well. In this model, we consider a lattice of spins. Each spin (an atom with a magnetic moment) is assumed to interact only with its nearest neighbour. The Hamiltonian for such a system is given by:

$$H = -\sum_i \sum_j J_{ij} S_i \cdot S_j. \tag{8.4}$$

In (8.4), J_{ij} is the strength of interaction between the spins (S_i and S_j) at the lattice sites i and j. The negative sign in (8.4) ensures that a parallel spin orientation corresponds to lower energy and is favoured (which is what happens in a ferromagnetic material). The summation over i calculates the Hamiltonian over the entire lattice, while the summation over j is a summation over the nearest neighbours of the ith spin. This implies that $i \neq j$, i.e. a spin cannot interact with itself. Another assumption of the Ising model is that a spin can take only one of two possible orientations (therefore, S_i can be either $+1$ or -1. These two directions can also be referred to as 'up' and 'down'. Physically, these two directions are parallel and antiparallel, respectively, with reference to an external magnetic field. The Hamiltonian in (8.4) describes the case in which there is no external magnetic field. In the presence of a magnetic field, there will be an additional term in (8.4) that represents the interaction between each spin and the external magnetic field. Even in the absence of an external magnetic field, we can still only talk about two possible orientations. In some systems, the spin can take any orientation within a plane (the XY model) or any orientation in three dimensions (the 3D Heisenberg model). We will discuss only the Ising model, which is the simplest possible model for magnetic systems.

At 0 K, the interaction between spins causes all spins to align in the same direction (figure 8.2), giving rise to the largest possible value for magnetisation. The system is now in a ferromagnetic phase. As the temperature increases, thermal fluctuations cause a few of the spins to flip (or point in the opposite direction to that of their neighbours), thereby reducing the magnetisation (figure 8.3). We can thus expect the magnetisation to decrease continuously as a function of temperature, so that the average magnetisation becomes zero at a particular temperature known as the Curie temperature. Above the Curie temperature, the system exhibits para-magnetism. The transition from the ferromagnetic to the paramagnetic phase is an example of a second-order transition. In a second-order transition, the Gibbs free energy and its first derivatives are continuous (entropy, magnetisation, and volume are the first derivatives of the Gibbs free energy defined by $G = U - TS + MH$ for magnetic systems).

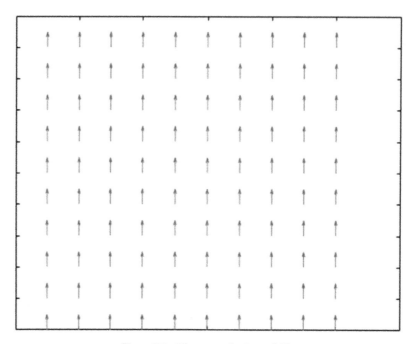

Figure 8.2. Alignment of spins at 0 K.

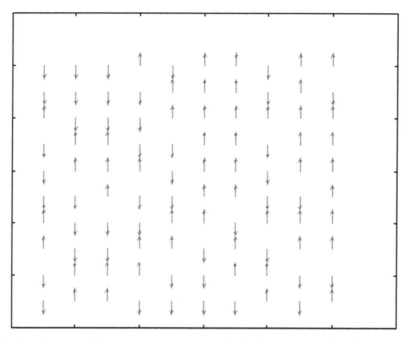

Figure 8.3. The alignment of spins at a temperature close to the Curie temperature. Note that there are almost equal numbers of up and down spins.

For non-magnetic systems, the term MH in the Gibbs free energy is replaced by PV.

Theoretically, the magnetisation M is expected to decrease with temperature according to the relation:

$$M \propto (T_c - T)^\beta. \tag{8.5}$$

Here, T_c is the Curie temperature and β is the critical exponent corresponding to the order parameter. The value of β depends on the model and also the dimension of the system.

It is interesting to study the Ising model in different dimensions:

- There is no transition to the ferromagnetic phase in 1D
- There is a transition in 2D, and the model can be solved analytically (the partition function and critical exponents can be evaluated)
- No analytical solution has been discovered to date for the 3D Ising model.
- In four or higher dimensions, the results of the Ising model are identical to those given by mean-field theories.

The critical exponent β is expected to be 1/8 for the 2D Ising model. For the 3D Ising model, β has been determined to be 0.32. To determine β, we need to simulate the Ising model. We can set up a configuration of a finite number of spins (placed on a lattice) on the computer. We would like to simulate an infinite lattice of spins, but since this is not possible on a computer with finite memory, we use periodic boundary conditions. Let us say our lattice is a square lattice and it has a size of $N \times N$. All spins in the interior of the lattice will have four nearest neighbours, while the spins along the four edges will only have three and those at the corners will only have two nearest neighbours. The presence of the boundary gives rise to edge effects, which broaden the transition and also modify the critical exponents. The use of periodic boundary conditions will reduce (but will not eliminate) the boundary effects. To ensure that all spins have four nearest neighbours, we treat the spins in the first column as neighbours of the spins in the last column (and vice-versa). A similar boundary condition is used for the first and last rows.

To simulate the Ising model and obtain the magnetisation as a function of temperature, we define an initial orientation for all the spins. All spins can point in the same direction at 0 K. This, of course, is the equilibrium configuration at 0 K that corresponds to the maximum magnetisation. However, what should be the initial configuration for $T > 0$ K? After all, we do not know the equilibrium configuration for any finite temperature. That is precisely what we are trying to determine.

We can adopt any random configuration as the initial configuration and use the Metropolis algorithm to determine the equilibrium configuration. The following MATLAB program uses the Metropolis algorithm to simulate the Ising model for a square lattice with 10×10 spins.

```
'Ising Model using Metropolis algorithm';
'zero field';
clear;
lattice=10;
'initializing the direction of all spins to point up';
for i=1:lattice
    for j=1:lattice
        u(i,j)=0;
        v(i,j)=1;
        x(i,j)=j;
        y(i,j)=i;
    end
end
m(1)=sum(sum(v));
quiver(x,y,u,v,0)
'initial T set to 1. Boltzmann constant and J are set to 1';
T=0.1;
J=1;
H=0;
z=1;
for i=2:lattice-1
    for j=2:lattice-1
        H=H-J*v(i,j)*(v(i-1,j)+v(i+1,j)+v(i,j-1)+v(i,j+1));
    end
end
' NO need to compute the energy of the entire lattice just
concentrate on the spin being flipped';
while T < 2.1

for k=1:1000
    r1=randi(lattice,1,1);
    r2=randi(lattice,1,1);
    'choosing a spin at random';
    'applying periodic boundary conditions';
    if r1==1
        r11=lattice;
    else
    r11=r1-1;
    end
    if r1==lattice
```

```
            r12=1;
        else
        r12=r1+1;
        end
   if r2==1
            r21=lattice;
        else
        r21=r2-1;
        end
        if r2==lattice
            r22=1;
        else
            r22=r2+1;
        end
        'dH is the change in the energy due to the spin flip';
        dH=2*J*v(r1,r2)*(v(r11,r2)+v(r12,r2)+v(r1,r21)+v(r1,r22));
        p=exp(-dH/T);
        r=rand;
        if p>r
            v(r1,r2)=-v(r1,r2);
        end
        clf;
     'calculating the magnetization m';
        m(k+1)=abs(sum(sum(v)));
    'Note that we take the absolute value of the magnetisation as a
    positive magnetisation is physically equivalent to a negative
    value';
        j=1:k+1;

end
avem(z)=mean(m(500:1000));
'leaving out the first 500 configurations as equilibrium would bot
have been reached';
T=T+0.1;
'incrementing the temperature';
z=z+1;
end
z1=0.1:0.1:T-0.1;
plot(z1,avem)
```

Note that at any given temperature, the system's configuration is changing all the time and what we are determining is the average magnetisation of many configurations. In a sense, we are determining the time average of the magnetisation, which is equal to the ensemble average by the ergodic hypothesis. The Metropolis algorithm has been used to determine the minimum energy of a system (see the discussion of simulated annealing in section 8.4). Note that the minimum energy always corresponds to the case in which all spins are aligned in the same direction, but according to the Boltzmann factor, $\exp(-E/kT)$, the probability of a high value of E increases as the temperature increases. Therefore, the probability of lower values of magnetisation increases as T increases, which is why the average magnetisation decreases with an increase in temperature.

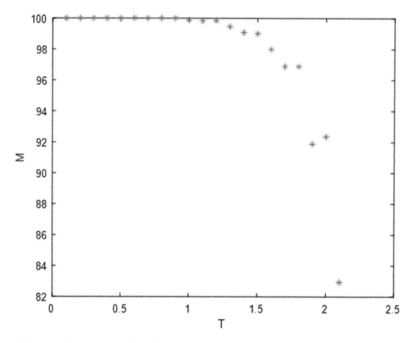

Figure 8.4. Magnetisation as a function of temperature for a 10×10 square lattice. We have set $J = k = 1$ to obtain this result.

Another way of understanding this is to realise that systems not only try to achieve the lowest energy state, but that entropy also has to be maximised (in accordance with the second law of thermodynamics). We say that the Gibbs free energy $G = U - TS + PV$ has to be minimised to account for both factors. Minimising the Gibb free energy minimises the internal energy U of the system and maximises the entropy S.

If we run the program given on the previous page, we get a plot of magnetisation as a function of temperature. Figure 8.4 is an output obtained from one run of the program. Note that the magnetisation as a function of temperature is not a smooth curve. This is because of the random nature of the Monte Carlo method.

Furthermore, the data points in figure 8.4 are too noisy to allow an accurate determination of the critical exponent β defined in (8.5). To obtain better results, the reader should modify the above program. Try modifying the program as follows:

- Increase the lattice size
- Average over a larger number of configurations
- Increase the temperature in smaller steps
- Modify the number of initial configurations to be neglected

For a correct determination of the critical exponents, the temperatures close to the critical temperature T_c are crucial. Hence, at temperatures very close to T_c, one has to use small steps. As one approaches the critical temperature, the dynamics of the system slow down (a phenomenon known as the 'critical slowing down')[1]. Methods have recently been formulated to overcome critical slowing down (see footnote 1 below).

How do we modify the procedure if we are trying to study an antiferromagnetic system? In an antiferromagnetic phase, the interaction between nearest neighbouring spins favours an antiparallel arrangement. Therefore, the interaction constant J in the Hamiltonian will be negative for an antiferromagnetic system, but this is subject to a complication, which is that the average magnetization of an antiferromagnet is zero, just like in the paramagnetic phase. To distinguish the two phases, we look at the magnetisation of a sublattice (which consists of only half the total number of spins). To understand this concept, consider the original lattice at 0 k to be like a chessboard. The white squares correspond to the 'up' spins, while the black squares correspond to the 'down' spins. At a finite temperature, some of the spins will have flipped. The sublattice corresponds to either the black squares or the white squares. The magnetisation of the sublattice exhibits a critical behaviour near the Neel temperature, just like a ferromagnet.

8.3 Percolation theory

Monte Carlo methods have been extensively used to study the phenomenon of percolation, which has a close relationship with phase transitions. Percolation in everyday language refers to the phenomenon of the percolation of a liquid (say coffee decoction) through a porous medium (a pile of coffee grounds). However, the scope of percolation theory extends far beyond the flow of liquids through porous media: problems ranging from the electrical properties of disordered systems to the spread of diseases and even rumours can be tackled using percolation theory[2].

Specifically, percolation theory can be used to study:
1. Metal–insulator transitions
2. The conductivity of a wire mesh
3. The spread of disease in a population
4. The behaviour of magnets diluted by non-magnetic impurities
5. The flow of oil through porous rock

Most problems in percolation involve disordered materials. However, from a theoretical perspective, it is easier to frame and solve problems formulated on a lattice. Hence, we will confine our discussions to percolation on a lattice. Even with this restriction, we can pose and answer many interesting questions, as this also ties in nicely with the fields of phase transitions and network theory.

---

[1] The simulation of the Ising model is discussed in great detail in *Monte Carlo Simulations in Statistical Physics*, Kurt Binder and Dieter W Heerman, 5th edn, Springer, Berlin, 2010. The book also discusses methods that can be used to overcome critical slowing down and all other aspects of simulating the Ising model.
[2] You can read *Physics and Geometry of Disorder*, A L Efros, Mir Publishers, Moscow, 1986, for an enjoyable introduction to percolation theory.

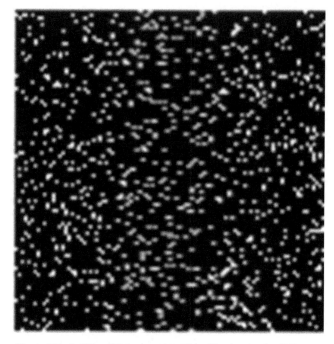

Figure 8.5. A 100 × 100 lattice with 10% of its sites in the ON state.

Consider a lattice (say a two-dimensional square lattice). Each site can be in one of two states (designated as either on/off or present/absent). Let a certain fraction x of the lattice be in the 'on' state (see figure 8.5). What is the minimum or threshold value of x (designated x_c) for which the 'flow' percolates from one edge of the square lattice to the opposite edge (say from the left edge to the right edge or from top to bottom)? Here, the flow could refer to the flow of electrons in a material or even the traffic flow through a network of roads.

We can think of another simpler system that will help us to understand the concept of percolation. Think of a two-dimensional lattice in which the lattice sites are filled with glass or steel balls. Initially, let all the sites have glass balls. The system will not allow current to flow from one side to the other. Now replace a certain fraction (x) of the glass balls with steel balls. What is the threshold value of x at which the system becomes a conductor? Here, we see that for small values of x, the system is an insulator, but beyond the threshold x_c, the system becomes a conductor. In other words, the system undergoes a phase transition from an insulative to a conductive state. Just as we defined critical exponents for the Ising model, we can define critical exponents for the percolation transition.

Below the threshold, isolated clusters of steel balls exist, but there is no spanning cluster (a cluster which extends from one end to the other). Above the threshold, a spanning cluster exists.

For low values of x, there is no spanning cluster of 'on' sites (some authors refer to them as occupied sites) that extends from the left to the right edge (or from top to bottom).

Figure 8.5 was generated by choosing a random number r for each lattice site. If r was greater than 0.1, the site was switched 'off' (these are also referred to as 'vacant' sites) and if it was less than 0.1, the site was switched 'on'. The probability p that r is less than 0.1 is, of course, 0.1, as the random number generator uses a uniform distribution. In this case, we notice that there is no spanning cluster. As we increase the value of p (and hence the fraction x of 'on' sites), we are likely to encounter the threshold value of p (p_c), above which a cluster of 'on' states exists that spans from the left to the right (or top to the bottom). Note that since the sites that are turned 'on' are randomly chosen, we need to repeat the simulation many times and determine the average value of p_c.

Figure 8.6 was generated with roughly 50% of its sites (selected randomly) in the 'on' state. You might now think that there is a possibility that at least one cluster of 'on' sites spans the lattice.

How do we write a program that determines the percolation threshold?

- Start with a 2D or 3D array in which a certain fraction of elements are 1 (filled sites), and the remaining elements are 0 (unfilled sites).
- Scan (read) the elements of the array, one by one, starting from the top left corner.
- If a non-zero element (site) is encountered, label it as 2.
- Look at the neighbours of this site and then the neighbours of the neighbours and label all of them (which have a value 1) as 2.

Figure 8.6. A 100×100 lattice with 50% of its sites in ON state.

- All the neighbours that have a value of 1 are added to an array so that we can later look at the neighbours of these pixels.
- Once all the 1s have been relabeled, we can check for end-to-end connectivity. This can easily be achieved by scanning the labelled array and checking whether the label 2 is found in the rightmost column (assuming that the spanning cluster is from left to right).

MATLAB has an inbuilt function, 'bwlabel', which labels a binary array. We can use this function and then check for end-to-end connectivity, as mentioned above.

The algorithm for labelling that we have discussed above is iterative in nature, i.e. we iteratively look at the neighbours of sites with a value of one. The algorithm can also be written recursively, i.e. we look at the neighbours of sites that have a value of one, then look at the neighbours of these neighbours, and so on.

The percolation threshold is similar to a phase transition. Just as we define critical exponents for a phase transition (recall the discussion of the Ising model in section 8.2), we can define various critical exponents for the percolation problem.

We can define a critical exponent for the mean connectedness length $\xi(p)$:

$$\xi(p) \sim |p - p_c|^{-\upsilon}. \tag{8.6}$$

One can also look at the fraction of occupied sites that are part of the spanning cluster:

$$P_\infty(p) = \frac{\text{number of sites in the spanning cluster}}{\text{total number of sites that are occupied}} \sim |p - p_c|^{\beta}. \tag{8.7}$$

The percolation discussed above is known as site percolation, since the lattice sites can be in one of two possible states. Bond percolation has also been studied extensively. In bond percolation, a random number is generated for each bond, which decides whether the bond will connect the two adjacent lattice sites or is cut. The critical exponents and percolation thresholds for bond percolation differ from the corresponding values for site percolation. This can be understood intuitively in the following manner: removing a lattice site is equivalent to removing all four bonds that connect that site to its neighbours on a square lattice.

8.4 Simulated annealing

We now present an application of the Metropolis algorithm that has gained wide currency in many areas far removed from physics. 'Annealing' refers to the process of slow cooling by which crystals are grown from a melt. It is well known that the periodic arrangement of the atoms in a crystal corresponds to the lowest energy state. It is more correct to say that a system of atoms chooses the state at a particular temperature and pressure that corresponds to the lowest Gibbs free energy (this minimises the system's internal energy and maximises the entropy). Imagine that you are trying to grow a crystal from the molten state of a material. You have to reduce the temperature gradually to form a perfect crystal. A sudden reduction in temperature will freeze the atoms in the positions in which they are found in the liquid state (which would, of course, be a disordered state). This gradual reduction in temperature is called annealing.

Just as the material needs to move to the lowest energy state, there are many problems in which we need to minimise a certain quantity. Hence, we can use simulated annealing, in which we slowly 'cool' a system so that it can move to the global minimum. 'Cooling' the system at a very fast rate would freeze it at a local minimum.

We now present a specific problem in order to understand the process of simulated annealing[3]. The 'travelling salesman' problem has been studied widely in computer science, as it is a problem that occurs in many situations.

Imagine a salesperson who has to visit many (say ten) cities (or localities within a city) as part of his/her job. The salesperson would like to plan his/her itinerary so as to minimise the total distance travelled. The brute force method of solving this problem is to compute the total distance travelled for each of the 10! possible permutations of cities (assuming that he/she starts from a particular city and returns to the same city after visiting every other city only once).

We know that computing the total distance travelled for each of the 10! possible routes (10! = 3628800) is too tedious, and in real-life situations, the number of cities could be larger. Note that the gradient descent method, which is widely used to solve optimisation problems, can get stuck in local minima.

A way forward is to use the Metropolis algorithm that we discussed in section 7.6. We start with a random route (i.e. cities arranged in a random order). A random initial configuration (order of cities) is shown in figure 8.7. We compute the total distance travelled for this configuration. The total distance travelled by the salesman will play the role of energy in this problem. We exchange the order of the visits to two of the cities to generate a slightly different configuration. We accept the new listing of cities (configuration) if the total distance travelled is reduced. If the total distance travelled is greater after the exchange, we do not immediately reject the new order but generate a random number r $(0 < r < 1)$.

If

$$r < w = \frac{\exp(-E_{\mathrm{trial}}/T)}{\exp(-E_i/T)}, \tag{8.8}$$

we accept the trial configuration that was generated. If (8.8) is not satisfied, we reject the trial configuration and move on to another trial configuration. Here, E_{trial} and E_i are the total distances travelled in the new and old configurations, respectively. Note that w in (8.8) is nothing more than the ratio of the Boltzmann factors with the Boltzmann constant set to one, as it is not relevant to this problem. The fictitious temperature T has the dimensions of a distance. We start the annealing by initially choosing a high value for the temperature T. This initial value can be larger than the largest estimate we can come up with for the total distance travelled.

---

[3] The original research article on simulated annealing, 'Optimization by simulated annealing', was written by S Kirkpatrick *et al*; see 1983 *Science* **220** (4598) 671–680. This article discusses many applications of simulated annealing, including the travelling salesman problem.

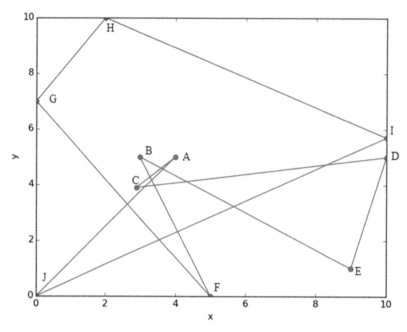

Figure 8.7. A possible configuration of cities for the travelling salesman problem. The total distance travelled in this case is 64.66.

We generate many trial configurations at each temperature and use (8.8) to accept or reject them. Once we have a sufficient number of configurations (i.e. we feel that the system has 'equilibrated' at that 'temperature'), we lower the temperature and repeat the process. Note that the process of simulated annealing is only useful if the total number of configurations that we generate for all temperatures is less than $n!$ (where n is the total number of cities).

Once the simulated annealing is implemented, if we start with the configuration shown in figure 8.7, we obtain a configuration (figure 8.8) that is very close to the absolute minimum.

To obtain the solution (figure 8.8), we started with an initial temperature of 200 (much greater than the distance of the random configuration initially chosen) and progressively reduced it to the temperature[4] shown in figure 8.8.

The moral of simulated annealing (and the Metropolis algorithm) is that to win a war, it is sometimes necessary to lose a battle!

---

[4] More details about the use of simulated annealing to solve the travelling salesman problem can be found in 'Efficiency of the simulated annealing algorithm in solving the traveling salesman problem', P S Jain and P K Thiruvikraman *Physical Science and Biophysics Journal* 2019 **3**(1) 000115.

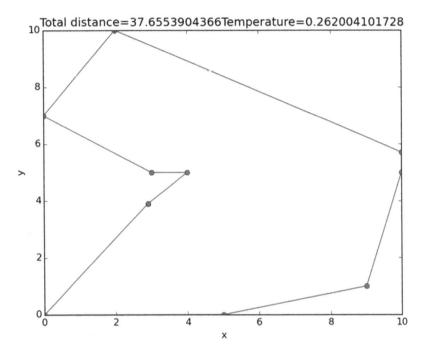

Figure 8.8. Solution to the travelling salesman problem obtained by simulated annealing.

Exercises

8.1. Modify the program for the Ising model (section 8.2) by: (a) increasing the size of the lattice, (b) the number of configurations over which you average to get the average magnetisation at a particular temperature, and (c) the number of initial configurations to be neglected.

Do the critical exponents and transition temperature vary with the variation in the parameters?

8.2. Modify the program for the Ising model to determine the critical exponents for an antiferromagnetic to paramagnetic phase transition.

8.3. In the travelling salesman problem, we assumed that straight lines could connect the cities that the salesman had to visit. In a real-life situation, he/she would have to travel along roads. Assume that the roads form a square grid. Do you think we need to consider multiple roads that connect two different cities as parts of different configurations? Why? In an even more realistic situation, the roads may not form a perfect square grid (this is approximately equivalent to blocking some of the roads on a square grid). Would you now need to consider different possible roads that connect two cities as part of two different configurations?

Chapter 9

Ordinary differential equations

9.1 Differential equations in physics

Differential equations lie at the heart of physics. Most of the equations used in physics are differential equations: consider, for example, Newton's second law, the Schrödinger equation, or Maxwell's equations. All of them are differential equations. In fact, most of them are partial differential equations. However, these partial differential equations reduce to ordinary differential equations under some circumstances.

Consider Newton's second law for motion in one dimension:

$$F_x = m\frac{d^2x}{dt^2}. \tag{9.1}$$

This is a second-order ordinary differential equation.

For the case of a simple pendulum oscillating with a small amplitude, Newton's second law reduces to:

$$m\ddot{\theta} = -mg\frac{\theta}{l}. \tag{9.2}$$

Similarly, the one-dimensional time-independent Schrödinger equation (9.3) is a second-order ordinary differential equation:

$$\frac{-\hbar^2}{2m}\frac{d^2\psi}{dx^2} + V(x)\psi = E\psi. \tag{9.3}$$

The important point to note about equations (9.2) and (9.3) and many other physical equations is that they are linear differential equations. One is therefore tempted to ask: is nature linear?

A little thought will lead to the conclusion that linearity is not inherent in nature. For instance, the equation for a simple pendulum is actually:

$$m\ddot{\theta} = -mg\frac{\sin \theta}{l}. \tag{9.4}$$

The right-hand side of (9.4) contains the sine function of the dependent variable, making the equation highly nonlinear. Equation (9.4) reduces to (9.2) after we impose the condition for small oscillations. Many other equations, for instance, the wave equation, are linearized by making assumptions similar to the small-angle approximation for the simple pendulum.

Physicists and mathematicians have traditionally devoted their time to solving linear equations simply because they are easier to solve and have some convenient properties that are absent from most nonlinear equations. For example, the principle of superposition, which is the cornerstone of optics, is true only for linear differential equations.

This tendency of mathematical physicists to confine their attention to linear differential equations could be one of the reasons why the field of nonlinear dynamics and chaos only developed in the second half of the 20th century. It turns out that most nonlinear differential equations can only be solved using numerical techniques. Therefore, the field of nonlinear dynamics had to await the development of high-speed computers. Nonlinearity was lurking around the corner all the time, but we required computers to explore it in its full glory. It was while the mathematician Ed Lorenz[1] was trying to solve a system of nonlinear differential equations that he chanced upon the phenomenon of chaos.

If we eliminate the constraint of linearity, many real-life problems can be modelled, and the corresponding differential equations can be solved numerically.

For instance, consider the case of projectile motion with air resistance. Projectile motion is studied by all students in introductory physics courses, but all the problems given in textbooks ask us to ignore the effects of air resistance. In real life, we cannot ignore the effects of air resistance. The following equation assumes that the force on the projectile due to air resistance is proportional to the square of its instantaneous velocity:

$$m\frac{d^2y}{dt^2} = -mg - k\left(\frac{dy}{dt}\right)^2. \tag{9.5}$$

Equation (9.5) can be solved by analytical means. Many nonlinear equations do have analytical solutions. It is sometimes convenient to solve nonlinear equations using numerical techniques because, even if an analytical solution is available, we finally need to evaluate the value of the function at specific points.

Solving (9.5) can tell us many things about the effect of air resistance. Will the trajectory still be a parabola? In the absence of air resistance, the trajectory is a parabola, as shown in figure 9.1.

However, in the presence of air resistance, the trajectory is not a parabola, as can be seen from figure 9.2.

---

[1] See chapter 9 of *Nonlinear Dynamics and Chaos* by Steven Stroking, Westview Press, Boulder, CO, 2000 for a detailed treatment and *Chaos: Making a New Science* by James Gleick, Viking Books, New York, 1987 for a popular account of chaos.

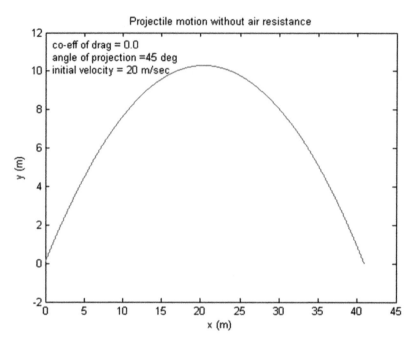

Figure 9.1. Trajectory of a projectile in the absence of air resistance.

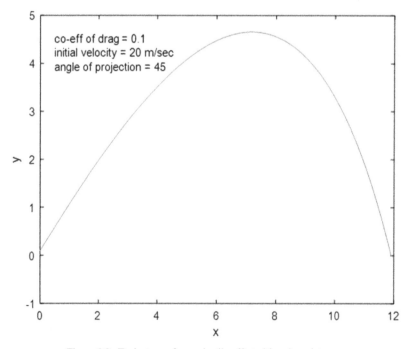

Figure 9.2. Trajectory of a projectile affected by air resistance.

Furthermore, we can ask whether the angle corresponding to the maximum horizontal range of a projectile is 45°. This is left as an exercise for the reader.

We can explore a wide range of problems if we use numerical techniques to solve differential equations. Introductory courses on quantum mechanics confine themselves to solving the Schrödinger equation for a few standard potentials: a particle in a box, the hydrogen atom, and harmonic oscillators. As we have seen in chapter 3 (section 3.10), we can obtain the energy levels of a finite square well potential using numerical methods. While that only involved root finding, we might need to solve the Schrödinger equation using numerical techniques for some other problems.

9.2 The simple Euler method

The Euler method is the simplest numerical technique for solving ordinary differential equations. Here, we consider initial-value problems, i.e. the initial position and velocity of a particle are known, and we need to determine the position and velocity of the particle at a later time. To begin with, let us consider first-order differential equations. Specifically, we are looking at a first-order ordinary differential equation of the form:

$$\frac{dy}{dx} = f(x, y).$$
(9.6)

We are trying to solve an initial-value problem, i.e. given $y(x_0)$, we are trying to determine $y(x)$. Consider the Taylor series expansion of $y(x)$ about $y(x_0)$:

$$y(x) = y(x_0) + (x - x_0)y'(x_0) + \frac{(x - x_0)^2}{2!}y''(x_0) + \dots$$
(9.7)

If we retain only the first two terms. we have the simple Euler method. In other words, we break down the interval from x_0 to x into small steps of size h. Within each interval, the function $f(x,y)$ is assumed to have a constant value (which is the value at the beginning of the interval). Note the similarity of this approximation to the rectangular approximation used in numerical integration (section 6.3).

Using the initial condition and a small step size for x, we can determine the value of y at a nearby location. Repeating this iteratively, we can determine the value of y for any value of x.

For example, consider the differential equation

$$\frac{dy}{dx} = 1 + y^2$$
(9.8)

with an initial condition of $y(0) = 0$.

A MATLAB program for the simple Euler method is shown below:

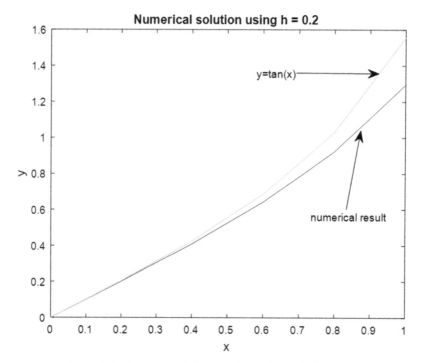

Figure 9.3. Comparison of the numerical and analytical results.

```
'Simple Euler  method';
y(0)=0;
h=0.2;
x=0:h:2;
for i=1:size(x,2)
    y(i+1)=y(i)+h*(1+y(i)^2);
end
figure
plot(x,y(1:size(x,2)),'r');
```

This differential equation can be solved by performing integration, and the solution is $y=\tan(x)$. A comparison between the numerical results obtained using the above program and the analytical results is shown in figure 9.3.

Note that in figure 9.3, the error in the numerical result, which is nothing more than the difference between the values obtained numerically and analytically, grows with increasing values of x. Of course, the magnitude of the error can be minimised using a smaller value of h (figure 9.4).

The error incurred using $h = 0.1$ can be minimised further using an even smaller value of h (figure 9.5).

The error in the simple Euler method is computed using (6.20), as this method uses the same approximation as the rectangular approximation. From (6.20), we see that

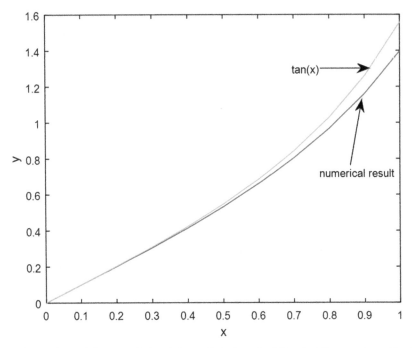

Figure 9.4. Comparison of the numerical and the analytical results. Note that the error is smaller than the error introduced by h=0.2.

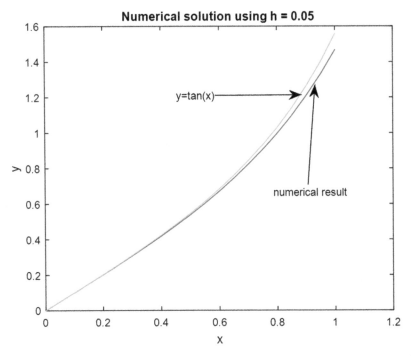

Figure 9.5. Numerical solution for h=0.05. Notice the reduction in error compared with the errors shown in figures 9.3 and 9.4.

the error in one step is proportional to h^2. If there are N steps between the initial point x_0 and the final point x, the total error will be proportional to $N h^2$. But $N = (x - x_0)/h$. Therefore, by substituting $N = (x - x_0)/h$ for $N h^2$, we can see that the total error is proportional to h.

The simple Euler method can also be used to solve second-order ordinary differential equations. We need to add one more step to each iteration. A physical example will help to clarify the procedure. Assume that you are given the initial position and velocity of a particle and have to find its subsequent position and velocity using Newton's second law. We discretize time into small intervals, Δt. During each interval, the forces acting on the particle are assumed to be constant. The forces could be a function of the position of the particle. For example, for motion due to a central force f (r), the force depends on the particle's position. However, if Δt is small, the force can be assumed to be constant within this interval. From this constant value of the force, calculated using the particle's position and velocity at the beginning of the interval, we can obtain its acceleration using Newton's second law. Using the acceleration, we can update its velocity during the next time instant and subsequently update its position using the new value of the velocity.

A program that computes the motion of a projectile subject to air resistance is shown below. The program uses the simple Euler method and follows the methodology outlined above. The output of this program is shown in figure 9.2.

```
'Projectile Motion';
v=input('enter initial velocity');
theta=input('enter initial angle of projectile with respect to
horizontal in degrees');
theta=theta*pi/180
vx=v*cos(theta)
vy=v*sin(theta)
v=(vx^2+vy^2)^0.5;
'k is the constant occuring in the air resistance';
k=input('Co-efficient of air resistance');
dt=0.0001
x=0;
y=0.1;
i=1;
while(y>0)
    fx=k*v*vx;
' The drag is proportional to v², but we need to resolve the drag
along the x and y axes'
    fy=k*v*vy;
    vx=vx-fx*dt;
    vy=vy-fy*dt-9.8*dt;
     v=(vx^2+vy^2)^0.5;
    x(i+1)=x(i)+vx*dt;
    y(i+1)=y(i)+vy*dt;
    i=i+1;
    plot(x,y);
    pause(0.01);
end
```

It is clear that using the value of the function at the beginning of the interval and ignoring its value at the end of the interval is the main source of error in the simple Euler method. This source of error is corrected in the modified and improved Euler methods.

9.3 The modified and improved Euler methods

In the modified Euler method, we use Euler's method to guess the solution at the midpoint. We then use this result to calculate the next half-step.

Therefore, in the modified Euler method, we have:

$$y\left(x_o + \frac{h}{2}\right) = y_o + \frac{h}{2}f(x_o, y_o). \qquad (9.9)$$

Equation (9.9) is nothing more than the use of the simple Euler method to obtain $y(x_0+h/2)$. Using the value obtained in (9.9), we can obtain the value of y at the end of the interval using the equation:

$$y(x_o + h) = y(x_o) + hf(x_{\mathrm{mid}}, y_{\mathrm{mid}}). \qquad (9.10)$$

The modified Euler method is not equivalent to the simple Euler method with a step size of $h/2$. If it were equivalent, then we would have used $y(x_{\mathrm{mid}})$ in (9.10) instead of $y(x_0)$.

The improved Euler method is implemented using the equation:

$$y(x_o + h) = y(x_o) + h\left[\frac{f(x_o, y_o) + f(x_o + h, y_o + hf_o)}{2}\right]. \qquad (9.11)$$

In the improved Euler method, we use the value of dy/dx at the end of the interval (obtained using the simple Euler method). Using the average of dy/dx at the beginning and end of the interval, we then obtain $y(x_0+h)$. Note from equations (9.9)–(9.11) that both the modified and improved Euler methods use the simple Euler method, but the results obtained by the modified and improved Euler methods are more accurate. Both these methods are equivalent to the trapezoidal rule for numerical integration.

To see the equivalence of the improved and modified Euler methods and the trapezoidal rule, note that the differential equation (9.8) can be written as an integral:

$$y(x_o + h) = y(x_0) + \int_{x_o}^{x_o+h} f(x, y)dx. \qquad (9.12)$$

The total error in both these methods is proportional to h^2 (see section 6.3). Their error will therefore be less than that of the simple Euler method for the same value of h (see figures 9.6 and 9.7). Hence, these methods are known as second-order Runge–Kutta methods. We discuss Runge–Kutta methods in the next section to elaborate this statement.

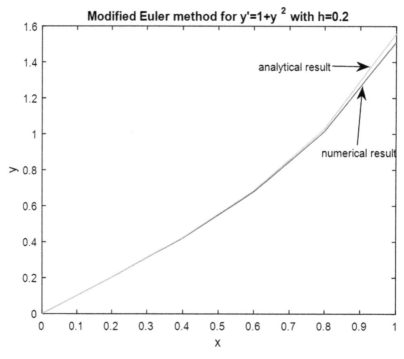

Figure 9.6. Results obtained using the modified Euler method. Compare these with figure 9.3.

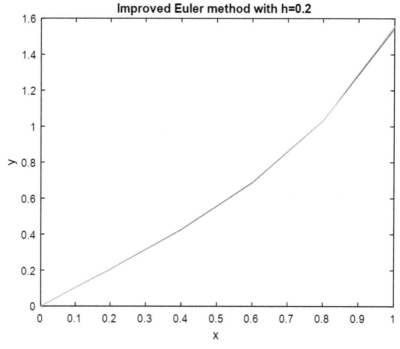

Figure 9.7. Improved Euler method. Compare these results with figures 9.3 and 9.6.

9.4 Runge–Kutta methods

All Euler methods can be written in the form:

$$y(x_o + h) = y(x_o) + h\left[\alpha f(x_o, y_o) + \beta f(x_o + \gamma h, y_o + \delta h f_o)\right]. \tag{9.13}$$

To justify this statement, consider the Taylor series expansion of $f(x,y)$:

$$\begin{aligned}
f(x, y) &= f(x_o, y_o) + (x - x_o)\frac{\partial f(x_o, y_o)}{\partial x} + (y - y_o)\frac{\partial f(x_o, y_o)}{\partial y} \\
&+ \frac{(x - x_o)^2}{2}\frac{\partial^2 f}{\partial x^2} + \frac{(y - y_o)^2}{2}\frac{\partial^2 f}{\partial y^2} + (x - x_o)(y - y_o)\frac{\partial^2 f}{\partial x \partial y} + \dots
\end{aligned} \tag{9.14}$$

Since we require $f(x_0 + \gamma h, y_0 + \delta h\, f_0)$ in (9.13), we evaluate it using the Taylor series:

$$f(x_o + \gamma h, y_o + \delta h f_o) = f(x_o, y_o) + \gamma h\frac{\partial f(x_o, y_o)}{\partial x} + \delta h f_o \frac{\partial f(x_o, y_o)}{\partial y} + \dots \tag{9.15}$$

Substituting for $f(x_0 + \gamma h, y_0 + \delta h\, f_0)$ from (9.15) in (9.13), we obtain:

$$y(x_o + h) = y(x_o) + h(\alpha + \beta)f(x_o, y_o) + h^2\beta\gamma\frac{\partial f}{\partial x} + h^2\beta\delta\frac{\partial f}{\partial y} + \dots \tag{9.16}$$

Compare (9.16) with the Taylor series expansion for $y(x_0+h)$:

$$\begin{aligned}
y(x_o + h) &= y(x_o) + hf(x_o, y_o) + \frac{h^2}{2}y''(x_o) + \dots \\
&= y(x_o) + hf(x_o, y_o) + \frac{h^2}{2}\left[\frac{\partial f(x_o, y_o)}{\partial x} + f(x_o, y_o)\frac{\partial f(x_o, y_o)}{\partial y}\right] + \dots
\end{aligned} \tag{9.17}$$

Comparing (9.17) and (9.16), we see that the following equations have to be satisfied:

$$a + b = 1 \tag{9.18}$$

$$\beta\gamma = 1 \tag{9.19}$$

$$\beta\delta = 1. \tag{9.20}$$

We can see that choosing $\alpha = \beta = 1/2$ and $\gamma = \delta = 1$ in (9.13) gives us the improved Euler method (9.11). Equation (9.13) also gives us a clue about how we can derive more accurate methods.

The detailed derivation of the expressions for the fourth-order Runge–Kutta method is rather involved, and we only state the result here.

The fourth-order Runge–Kutta method is based on the following equations:

$$f_0 = f(x_o, y_o) \tag{9.21}$$

$$f_1 = f\left(x_o + \frac{h}{2}, y_o + \frac{h}{2}f_o\right) \tag{9.22}$$

$$f_2 = f\left(x_o + \frac{h}{2}, y_o + \frac{h}{2}f_1\right) \tag{9.23}$$

$$f_3 = f(x_o + h, y_o + hf_2). \tag{9.24}$$

Using equations (9.21)–(9.24), we can obtain $y(x_0+h)$ from $y(x_0)$ using the equation:

$$y(x_o + h) = y(x_o) + \frac{h}{6}\left[f_o + 2f_1 + 2f_2 + f_3\right]. \tag{9.25}$$

The fourth-order Runge–Kutta method is derived using (9.13), but considering the Taylor series leads to a set of 13 equations with 11 unknowns. Choosing two of the unknowns leads to (9.25). The fourth-order Runge–Kutta method is similar to Simpson's rules for numerical integration, though you will notice some differences between (9.25) and (6.32), both of which use three points. The total error in the fourth-order method is proportional to h^4, just as in the case of Simpson's rules. This is the reason it is called the fourth-order Runge–Kutta method.

We can decrease the error in the Runge–Kutta methods even more using either (or both) of the following methods:

(i) Implement the Runge–Kutta method of a certain order for different step sizes (values of h) and use a technique similar to Richardson extrapolation (section 6.2).

(ii) Implement the Runge–Kutta method for two different orders. This modification is known as Runge–Kutta–Fehlberg method.

If we use the fourth Runge–Kutta method and obtain the value of y using step sizes h and $h/2$, then we have:

$$y_{\text{exact}} = y_h + kh^4. \tag{9.26}$$

The second term in the right-hand side of (9.26) is the error in the fourth-order Runge–Kutta method. We now repeat the computation using a step size of $h/2$ to obtain:

$$y_{\text{exact}} = y_{h/2} + k\left(\frac{h}{2}\right)^4. \tag{9.27}$$

Subtracting (9.27) from (9.26), we have:

$$0 = y_h - y_{h/2} + \frac{15}{16}kh^4. \tag{9.28}$$

Rearranging the terms in (9.28) gives us an expression for k:

$$kh^4 = \frac{16}{15}(y_{h/2} - y_h). \tag{9.29}$$

By substituting (9.29) into (9.26), we obtain:

$$y_{\text{exact}} = y_h + \frac{16}{15}(y_{h/2} - y_h) = \frac{16y_{h/2} - y_h}{15}. \tag{9.30}$$

Equation (9.30) does not actually give us y_{exact}, but it gives us value that is more accurate than that obtained using the fourth-order Runge–Kutta method.

In the Runge–Kutta–Fehlberg method, we compute the value of $y(x_0+h)$ using Runge–Kutta methods of two different orders and use a technique similar to the Richardson extrapolation.

We know that for the nth order Runge–Kutta method, the error of one step is proportional to h^{n+1}:

$$y_n(x_o + h) = y_{\text{exact}} + kh^{n+1}. \tag{9.31}$$

Similarly, for the Runge–Kutta method of the $(n+1)$th order:

$$y_{n+1}(x_o + h) = y_{\text{exact}} + kh^{n+2}. \tag{9.32}$$

From (9.31) and (9.32), we have:

$$k = \frac{y_n - y_{n+1}}{h^{n+1}}. \tag{9.33}$$

We can use this expression for k to calculate the step size required to reduce the error (or maintain it) below some threshold, ε:

$$kh_{\text{new}}^{n+1} = \frac{y_n - y_{n+1}}{h^{n+1}}h_{\text{new}}^{n+1} \leqslant \varepsilon. \tag{9.34}$$

The corresponding algorithm works as follows:
1. Calculate $y_n(x_o+h)$ and $y_{n+1}(x_o+h)$ from $y(x_o)$.
2. Calculate h_{new}.
3. If h_{new} is less than h, reject the propagation to x_o+h, redefine h and repeat step one.
4. If h_{new} is greater than h, accept the propagation to x_o+h, redefine h, and repeat step one to continue propagating the solution.

MATLAB has a built-in function, ode45, which uses Runge–Kutta methods of different orders to solve ordinary differential equations.

We close this section by providing some tips for getting accurate results using multistep methods.

Tips for obtaining correct results:
- Keep the time step small in relation to timescales involved in the problem.
- Check that the results do not vary with a change in the time step used (if they do, then reduce the time step).

- Monitor physical parameters that are supposed to be constant. For conservative systems, total energy should be a constant.
- Check the program for regimes in which an analytical solution exists.

9.5 The Taylor series method

The Euler and Runge–Kutta methods are multistep methods that use the Taylor series. There is also a single-step method that directly uses the Taylor series. We can write the Taylor series expansion for $y(x_0+h)$ as:

$$y(x_o + h) = y(x_o) + hy'(x_o) + \frac{h^2}{2!}y''(x_o) + \frac{h^3}{3!}y'''(x_o) + \dots \tag{9.35}$$

From (9.6), we have:

$$y'(x_0) = \frac{dy}{dx}\bigg|_{x_0} = f(x_0, y_0). \tag{9.36}$$

The higher-order derivatives that occur in (9.35) can be obtained by repeated differentiation of (9.6). We exemplify the procedure using the differential equation that we discussed earlier (9.8).

Example 9.1

Solve the differential equation $\frac{dy}{dx} = 1 + y^2$ using the Taylor series method. Find y (1) given that $y(0) = 0$. Use terms up to the second order.

Answer: Calculate the higher derivatives by hand and use the initial condition to get their numerical values. Differentiating the given equation gives us the second derivative:

$$\frac{d^2y}{dx^2} = 2y\frac{dy}{dx} = 2y(1 + y^2). \tag{9.37}$$

By substituting the initial value of y, we get $\frac{d^2y}{dx^2} = 2y(1 + y^2) = 0$ at $x = 0$. Proceeding further, we get the value of the third and fourth derivatives:

$$\frac{d^3y}{dx^3} = 2y'(1 + y^2) + 4yy' = 2 \text{ at } x = 0 \tag{9.38}$$

$$\frac{d^4y}{dx^4} = 2y''(1 + y^2) + 4yy'^2 + 4y'^2 + 4yy'' = 4 \text{ at } x = 0. \tag{9.39}$$

By substituting these into the Taylor series expansion about $y(0)$, we obtain:

$$y(1) = y(0) + y'(0) + \frac{1}{2!}y''(0) + \frac{1}{3!}y'''(0) + \frac{1}{4!}y''''(0) = 1 + \frac{2}{6} + \frac{1}{6} = 1.5. \tag{9.40}$$

The exact value of $y(1)$ is $\tan(1) = 1.5574$.

To achieve greater accuracy, we can use the Taylor series method as a multistep process in which we Taylor expand about the point at the beginning of the interval.

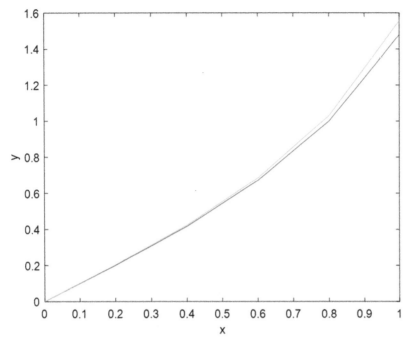

Figure 9.8. Multistep Taylor series method with a step size of 0.2 Compare these results with the results shown in figures 9.6 and 9.7.

Using the Taylor series method as a multistep process with terms up to the second derivative only, we get the results shown in figure 9.8, for which we used a step size of 0.2.

The differential equation solved in example 9.1 is a nonlinear equation. For linear equations, a recurrence relation can be derived that relates the derivatives of different orders, which makes it possible to easily compute higher-order derivatives (without working them out on paper!), as the following example shows.

Example 9.2

Determine $y(1)$ given that $y(0) = -1$, where y satisfies the differential equation $\frac{dy}{dx} = -2x - y$.

Answer: We compute the higher-order derivatives from the given differential equation:

$$\frac{d^2y}{dx^2} = -2 - y' \quad \frac{d^3y}{dx^3} = -y''. \tag{9.41}$$

In fact, for all derivatives higher than the second derivative, the recurrence relation is:

$$\frac{d^ny}{dx^n} = -\frac{d^{n-1}y}{dx^{n-1}}. \tag{9.42}$$

By substituting the values of the derivatives into the Taylor series, we obtain:

$$y(1) = y(0) + y'(0) + \frac{1}{2!}y''(0) + \frac{1}{3!}y'''(0) + \frac{1}{4!}y''''(0)$$

$$= -1+1 - \frac{1}{2!} + \frac{1}{3!} - \frac{1}{4!} + .. = -e^{-1} = -0.3679.$$

(9.43)

Of course, this exactly matches the solution obtained by the usual analytical methods, as we have taken an infinite number of terms in (9.43).

9.6 The shooting method

We close the discussion of ordinary differential equations by discussing a method for solving boundary-value problems. All the methods discussed so far in this chapter are geared towards solving initial-value problems. However, in certain situations, the solutions to differential equations have to satisfy some boundary conditions.

In the case of initial-value problems, we start from the initial point (or time) and evolve the system using the differential equation and the initial conditions to determine its subsequent evolution.

How do we use the methods discussed so far to solve boundary-value problems? We come across boundary-value problems (for instance) in quantum mechanics, where the function has to satisfy certain boundary conditions. The wavefunction (see section 3.10) has to vanish at the boundaries for some situations or has to definitely vanish at infinity.

Assume that you have to solve a second-order differential equation, say,

$$\frac{d^2u}{dx^2} = -\frac{u}{4},$$

(9.44)

subject to the boundary conditions: $u(0) = 0$ and $u(\pi) = 2$.

If this were an initial-value problem, we would use $u(0) = 0$ and the derivative of u at $x = 0$ as initial conditions to evolve the systems using the Euler or Runge–Kutta methods. Instead of the derivative, we now have the value of u at some other point. Since we do not know the derivative of u at the initial point, we simply guess its value and evolve the system using the Euler method (for example). Using the Euler method, we can compute the value of u at the endpoint ($x = \pi$). If we find that we have overshot the target $u(\pi) = 2$, then we can reduce the initial guess for the derivative.

In general, let u'_1 and u'_2 be the two guesses for the value of the derivative and let u_{b1} and u_{b2} be the corresponding values of u at the endpoint b. If u'_c is the value of the derivative that would produce the required value of u (u_p) at the boundary point b, then we have:

$$\frac{u'_c - u'_1}{u_b - u_{b1}} = \frac{u'_2 - u'_1}{u_{b2} - u_{b1}}.$$

(9.45)

Equation (9.45) assumes that changes in the value of the derivative cause proportionate changes in the value of the function at the endpoint.

Equation (9.45) can be rewritten to obtain the value of the derivative that would produce the required value of u at the boundary point. From (9.45):

$$u'_c = u'_1 + \left(\frac{u'_2 - u'_1}{u_{b2} - u_{b1}}\right)(u_b - u_{b1}). \qquad (9.46)$$

The MATLAB program shown below implements the shooting method:

```
'shooting method';
clear;
clf;
y(1)=0;
h=0.01;
x=0:h:pi;
dr=2;
v=0.1;
g1=v;
for i=2:size(x,2)
    a=-y(i-1)/4;
    v=v+a*h;
    y(i)=y(i-1)+v*h;
end
r1=y(size(y,2));
y1=y;
v=0.2;
g2=v;
for i=2:size(x,2)
    a=-y(i-1)/4;
    v=v+a*h;
    y(i)=y(i-1)+v*h;
end
r2=y(size(y,2));
y2=y;
dg=(dr-r2)*(g2-g1)/(r2-r1)+g2;   'implementation of (9.46)';

v=dg;
for i=2:size(x,2)
    a=-y(i-1)/4;
    v=v+a*h;
    y(i)=y(i-1)+v*h;
end
plot(x,y,'g',x,y1,'b',x,y2,'r');
```

Figure 9.9 shows the output of the program. Initially we chose 0.1 as the value of the derivative of u at $x = 0$; however, since we fell far short of the target at $x = \pi$, we increased the derivative to 0.2. Using these two values, we computed the required value of the derivative to be one. Using this value of the derivative gives the solution to the differential equation (green curve in figure 9.9) that also satisfies the required boundary condition.

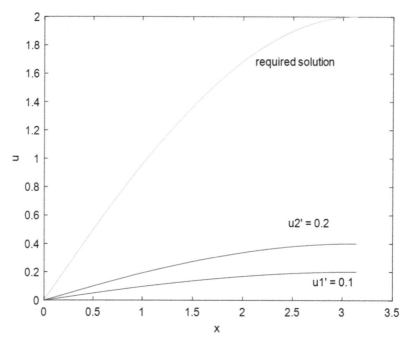

Figure 9.9. Results for the shooting method.

9.7 Applications to physical systems

So far, this chapter has discussed numerical techniques used to solve ordinary differential equations. When we consider many physical systems, it is more difficult to derive the differential equations than to solve them! This is because we need to first model the system, decide which parameters should be included in the model, and leave out those that we consider to be irrelevant. We have given one example of modelling a physical system in section 9.2. There, it was fairly obvious that we needed to consider gravity and air resistance as forces acting on the projectile and we then used Newton's second law to determine the trajectory of the projectile.

Another example that occurs in introductory mechanics is that of the simple pendulum. The equation of motion of a simple pendulum is given by:

$$ml\frac{d^2\theta}{dt^2} + mg \sin \theta = 0. \tag{9.47}$$

This can be written as:

$$\frac{d^2\theta}{dt^2} + \frac{g}{l} \sin \theta = 0. \tag{9.48}$$

Equation (9.48) is highly nonlinear because of the term containing $\sin\theta$. We would normally linearize it by saying that $\sin\theta \sim \theta$ for small amplitudes. But suppose we are interested in the behaviour of a simple pendulum oscillating with a large amplitude; we then have to solve (9.48) using an Euler or Runge–Kutta method.

A program that solves (9.48) using the simple Euler method is given below:

A program for solving (9.48) using the simple Euler method is given below:

```
'Program for large amplitude simple pendulum' (Euler's Method);

qo =1.5;
q(1)=qo;
w(1)=0;
iterations=10000
for i=2:iterations
a (i)=-sin(q(i-1));
w(i)=w(i-1)+a(i)*0.01;
q(i)=q(i-1)+w(i)*0.01;
end
i=1:iterations;
plot(i,theta);
```

In the above program, we have taken $g = l$.

Using the plot produced by the program, we can determine the time period for oscillation. Running the program for different amplitudes gives us the time period as a function of amplitude (figure 9.10).

Can you fit a polynomial to the data points in figure 9.10? Note that the constant term in the polynomial is obviously 2π, which is the period of a simple pendulum with $l = g$. Fitting a polynomial is left as an exercise for the reader (see exercises 9.6 and 9.7 at the end of the chapter). An automatic determination of the period from the output data of the above program is an interesting exercise. One way of achieving this is to look at the product of the values of w (angular velocity) in successive iterations. The product will change sign at the endpoints. The program

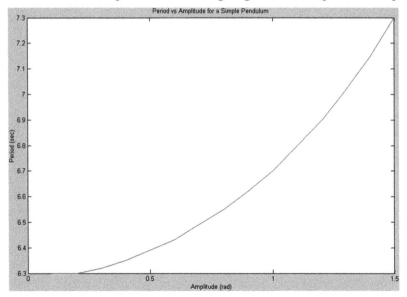

Figure 9.10. Time period as a function of amplitude for a simple pendulum.

can record the time corresponding to a change of sign in the product. The interval between two such instants will be equal to $T/2$.

The following example illustrate the modelling of physical processes which give rise to differential equations. In many cases, the differential equations can only be solved by numerical methods.

Example 9.3

Model the operation of a cannon and devise a differential equation that can be used to predict the velocity of the cannonball as it leaves the cannon.

Answer: The cannon consists of a long hollow cylinder. A gunpowder explosion in the cannon produces hot gases. The gunpowder is enclosed in a small chamber within the cannon. The chamber is closed on one side by the cannonball. The hot gases exert a pressure that is much greater than the atmospheric pressure on the other side of the cannonball. This difference in pressure exerts a net force on the cannonball, which is accelerated along the length of the cannon. The cannonball is accelerated until it leaves the cannon.

The velocity with which the cannonball exits the cannon depends on the length of the cannon and the excess pressure inside the cannon. Let us assume that we know the energy released during the explosion of the gunpowder. From this energy, we can work out the temperature and the pressure of the gas inside the cannon just after the explosion.

It is clear that the force exerted on the cannonball is not constant. The pressure inside the chamber drops as the gas expands (due to the motion of the cannonball). Heat loss will also contribute to a slight drop in pressure, but we can ignore this, as it is less important. The expansion of the gas is actually a non-equilibrious process, but if we ignore the heat losses, we can approximate the process using an adiabatic expansion.

An adiabatic process satisfies the following equation:

$$P_0 V_0^\gamma = P V^\gamma, \tag{9.47}$$

where P_0 and V_0 are the pressure and volume, respectively, of the hot gas immediately after the explosion and P and V are its pressure and volume, respectively, at a later time t. If A is the cross-sectional area of the cannon and x is the displacement of the cannonball after time t, we have from (9.47):

$$F = PA = \frac{P_0 V_0^\gamma A}{(V_0 + Ax)^\gamma}. \tag{9.48}$$

Equation (9.48) gives us the force acting on the cannonball at time t; therefore, the differential equation for the motion of the cannonball is:

$$\frac{d^2x}{dt^2} = \frac{P_0 V_0^\gamma A}{m(V_0 + Ax)^\gamma}. \tag{9.49}$$

Equation (9.49) can be solved numerically to obtain the instantaneous position and the velocity of the cannonball until it exits the cannon, i.e. when $x = L$ (the length of the cannon).

Example 9.4

A spring–mass system is placed beneath a rectangular box, as shown in figure 9.11. The box contains sand and has a long narrow slit at the bottom.

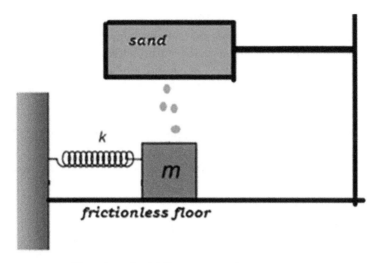

Figure 9.11. Sand falling on to a spring–mass system.

The block is set into oscillation and the slit is simultaneously opened. Initially, the mass of the block is m_0, but its mass increases linearly with time, as sand falls at a constant rate of C kg s^{-1} and sticks to the block. Assume that the spring is massless and that sand does not stick to the spring. Ignore air resistance.

(a) Write down the equation of motion for the block.
(b) If the spring constant $k = 10$ N m^{-1}, $m_0 = 1$ kg, and $C = 0.1$ kg s^{-1}, show that the motion is similar to that of a damped harmonic oscillator. What will be the nature of the oscillation (lightly damped/heavily damped/critically damped/undamped) to begin with? Explain your answer using relevant calculations.
(c) Does the nature of the damping change with time? Explain your answer.
(d) What will be the angular frequency of the system (to begin with)? Will the angular frequency change with time? If so, how and why?

Answer:
(a) The equation of motion for the block is:

$$m\frac{d^2x}{dt^2} = -kx - v\frac{dm}{dt}. \tag{9.50}$$

Equation (9.50) is also known as the 'rocket equation', as the equation of motion for a rocket is similar to this. In fact, all variable mass problems can be solved using the rocket equation. The rocket equation follows from an application of the conservation of momentum to the given system[2].

It is given that $dm/dt = c$; therefore, this equation can be rewritten as: $m\frac{d^2x}{dt^2} + c\frac{dx}{dt} + kx = 0$. This is the equation for a damped simple harmonic oscillator (but with a variable mass).

---

[2] See, for instance, *An Introduction to Mechanics* by Kleppner and Kolenkow, McGraw Hill, New York, 2009 for a discussion of the rocket equation.

(b) Here, $\gamma = c/m = 0.1$ and $\omega_0^2 = 100$.

Since $\omega_0^2 \gg \gamma^2 4$, the mass is subjected to light damping.

(c) Both the damping constant and the natural frequency change as the mass changes with time:

$$\omega^2 = \frac{k}{m_o + ct} \quad \frac{\gamma^2}{4} = \frac{c^2}{4(m_o + ct)^2}.$$

(9.51)

Since $\gamma^2/4$ decreases at a faster rate than ω^2, the nature of the oscillation will continue to be lightly damped. The damping decreases with time.

(d) Initially, $\omega_o^2 = 10$, but since the mass increases linearly with time, the natural frequency will decrease.

$$\omega(t) = \sqrt{\frac{k}{m_o + ct} - \frac{c^2}{4(m_o + ct)^2}}.$$

(9.52)

The above variation of the angular frequency is an approximation, because in the equation for the damped harmonic oscillator, the mass is a constant, whereas the mass here is a variable. Therefore, using the results obtained for the damped harmonic oscillator may not be completely correct. In fact, the decay of the amplitude is not exponential.

We can obtain the correct behaviour of the angular frequency using the Euler method (for example) to numerically solve (9.50).

The following program MATLAB program can be used to obtain the numerical solution:

```
clf;
clear;
mo=1;
c=1;
k=10;
x(1)=0.1;
x1(1)=0.1;
v(1)=0;
dt=0.0001;
t(1)=0;
for i=1:1000000
    m(i)=mo+c*i*dt;
    w(i)=(k/m(i)-c^2/(4*m(i)^2))^0.5;
    a=-(k/m(i))*x(i)-v(i)*c/m(i);
    v(i+1)=v(i)+a*dt;
    x(i+1)=x(i)+v(i+1)*dt;
    x1(i+1)=0.1*exp(-c*t(i)/2)*cos(w(i)*t(i));
    t(i+1)=t(i)+dt;
end

i=1:size(x,2);

plot(t,x,'r',t,x1,'b')
```

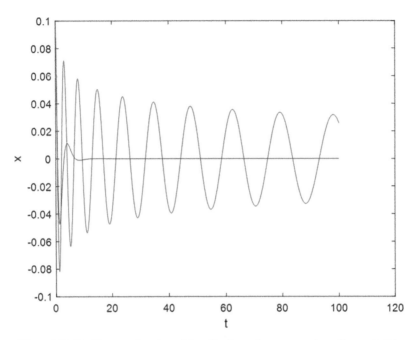

Figure 9.12. Position as a function of time for the spring–mass system of example 9.4.

The program also plots the exponential decay (blue curve) characteristic of the damped harmonic oscillator. Figure 9.12 shows that the numerical result (red curve) is very different from the exponential decay of the amplitude one might naively expect. This shows that numerical solutions are important in situations in which our intuition may sometimes lead us astray! Figure 9.12 does, however, confirm the increase in frequency with time.

Exercises

9.1 Consider the initial-value problem $\frac{dx}{dt} = 1 + \frac{x}{t}$, where $x(t=1) = 1$. Suppose you are trying to use the Taylor series method to calculate $x(t=2)$.
  (a) Write down the expression for $x(t=2)$ as a Taylor series expansion about $x(t=1)$.
  (b) Derive a general expression for the nth derivative of x with respect to t. In addition, indicate the minimum value of n for which this general expression is valid.
  (c) Simplify the Taylor series using (b).
9.2 Can the shooting method be used to solve a boundary-value problem in two dimensions? Why?
9.3 A quantity x satisfies the differential equation $\frac{dx}{dt} = \sin t - x$. You are given the initial condition $x(\pi) = 1$.

(a) Determine the value of x at $t = 2\pi$ using the Taylor series method. Use derivatives up to the fourth derivative to determine the result.

(b) Construct at a general expression for the nth derivative.

(c) Use the general relation and the Taylor series to determine (approximately) the number of terms you have to use for the error in the result to be less than one.

9.4 The program below is an implementation of the simple Euler method. There is only one mistake (a logical error, not a syntax error) in this program, due to which, it will not give the correct output. Identify the line number which has the error.

```
Line no:

1      y(1)=0.01;

2      h=0.1;

3      x=0:h:10;

4      for i=1:size(x,2)

5 y(i+1)=y(1)+h*(y(i)-y(i)^3);

6   end
```

9.5 Show that the modified Euler method agrees with the Taylor series for terms up to the order of h^2.

9.6 Figure 9.10 gives the time period of a simple pendulum as a function of amplitude. Extract a few data points from the figure. You may use the MATLAB function 'polyfit' to determine the coefficients of the polynomial which fits the given data.

9.7 Modify the program given for solving the equation of a simple pendulum so that the program automatically calculates the period of oscillation. Then use the polyfit command as mentioned in the previous problem.

9.8 A block of mass m is placed on a flat table and connected to a horizontal spring (the other end of the spring is attached to a wall). The coefficient of friction between the block and the table is μ. Write down (no need to solve) the differential equation which describes the motion of the block. Neglect viscous drag in this case.

9.9 Three dogs A, B, and C are at the three vertices of an equilateral triangle. At the stroke of a bell, they start running towards the instantaneous position of the next nearest dog, i.e. dog A runs towards dog B, B runs towards C, and C runs towards A. All the dogs run at the same constant speed. Simulate the subsequent motion of the dogs using the Euler method and give your conclusions.

IOP Publishing

Computational Methods Using MATLAB®
An introduction for physicists
P K Thiruvikraman

Chapter 10

Partial differential equations

10.1 Partial differential equations in physics

As pointed out in chapter 9, most of the equations of physics are differential equations. The heat/diffusion equation, the Laplace equation from electrostatics, and the time-dependent Schrödinger equation are examples of partial differential equations (PDEs) that occur in different branches of physics.

To solve partial differential equations, we use boundary conditions. For instance, solving the Laplace equation: $\nabla^2 V = 0$, involves determining the potential inside a region using a knowledge of the potential at the boundaries of that region. Hence, to solve a PDE, we have to solve a boundary-value problem. Most PDEs which are of physical interest are second-order linear equations. The finite difference method is one of the ways of solving partial differential equations. In the finite difference method, we replace the derivatives that occur in the PDEs with finite differences, using the expressions for the derivatives derived in section 6.1.

10.2 Finite difference method for solving ordinary differential equations

We will first apply the finite difference method to solve an ordinary differential equation (ODE); the same technique can then be applied to PDEs.

Consider the equation:

$$\frac{d^2y}{dx^2} = -\frac{y}{4}.$$

(10.1)

We wish to solve this equation subject to the boundary conditions:

$$y(0) = 0 \quad \text{and} \quad y(\pi) = 2.$$

This equation can be solved analytically (the solution is sinusoidal), but how would we solve it using the finite difference method? The first step is to replace the derivative(s) which occur in the equation with finite differences. To do this, we have to take the interval under consideration ($[0,\pi]$) and divide it into small intervals of width h.

From (6.7), we have the expression for the second derivative: $f''(x_1) = \frac{f(x_2)+f(x_0)-2f(x_1)}{h^2}$. By inserting this expression into (10.1), we obtain:

$$\frac{d^2y}{dx^2} = \frac{y(x_{i+1}) + y(x_{i-1}) - 2y(x_i)}{h^2} = -\frac{y(x_i)}{4}. \tag{10.2}$$

We can write (10.2) to obtain an expression for $y(x_i)$ in terms of the value of y at its neighbouring points:

$$y(x_i) = \frac{4(y(x_{i+1}) + y(x_{i-1}))}{(8 - h^2)}. \tag{10.3}$$

Equation (10.3) may not appear useful at first sight. The value of the unknown function $y(x_i)$ is given in terms of its value at two neighbouring points. If we divide the interval $[0,\pi]$ into steps of, say, 0.1, then except for the first and last points (the ends of the interval) the function is unknown at all other points. As with any iterative procedure, we assume the value of y at all points except the two ends of the interval. We then iteratively use (10.3) to determine the value of y at the interior points. After many iterations, the information about the boundary value will 'seep' into the region, subject to the constraint posed by the PDE, to converge to the stable (correct) solution.

The program shown below is an implementation of the finite difference method to solve (10.1). Here, we have assumed that y initially varies linearly with x. This initial variation will change on the iterative application of (10.3) and converge to the sinusoidal variation.

```
'finite difference method';

clear;
y(1)=0;
h=0.1;
x=0:h:pi;
error=10;
y(size(x,2))=2;
y2=2*sin(x/2);
'Choosing some initial values for y. Here we have chosen a linear
relation between y and x';
for i=1:size(x,2)
    y(i)=(2/pi)*x(i);
end

while error >0.0001
        clf;
        error=0;
for i=2:size(x,2)-1
    y1(i)=4*(y(i+1)+y(i-1))/(8-h^2);
    error=error+abs(y(i)-y1(i));
end
y=y1;
y(size(x,2))=2;
y(1)=0;
plot(x(1:size(y,2)),y,'b',x,y2,'r');
pause(0.01);
end
```

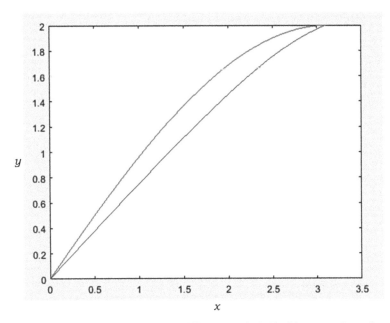

Figure 10.1. Output of the program for the finite difference method. The blue curve shows the result of the finite difference method after a few iterations, and the red curve is the analytical solution.

The output of the above program is shown in figure 10.1 (after a few iterations) and after many iterations in figure 10.2. Figure 10.2 shows that the output of the finite difference method converges to that of the analytical method after many iterations. If you run the above program, you can see the slow convergence of the numerical result to the analytical result in the MATLAB figure window. The pause command has been used to slow down the rate at which the convergence happens, so that you can easily view the changes in the figure window.

10.3 Finite difference method for solving PDEs

We first consider the Laplace equation in two dimensions and attempt to solve it using the finite difference method. The Laplace equation in two dimensions is given by:

$$\nabla^2 V = \frac{\partial^2 V}{\partial x^2} + \frac{\partial^2 V}{\partial y^2} = 0. \tag{10.4}$$

Using the finite difference expression (6.7) for the second derivatives, the above equation can be rewritten as:

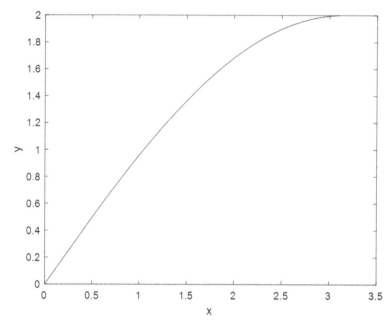

Figure 10.2. The numerical solution converges to the analytical result after many iterations. The red and blue curves are on top of each other.

$$\frac{V(x + h, y) + V(x - h, y) - 2V(x, y)}{h^2}$$
$$+ \frac{V(x, y + h) + V(x, y - h) - 2V(x, y)}{h^2} = 0. \tag{10.5}$$

Equation (10.5) can be rewritten as:

$$V(x, y) = \frac{V(x + h, y) + V(x - h, y) + V(x, y + h) + V(x, y - h)}{4}. \tag{10.6}$$

Equation (10.6) has a very simple interpretation. It shows that the value of the potential at a point (x,y) is the average of the potentials at its four nearest neighbouring points on a rectangular grid. In fact, this result holds for the Laplace equation in any dimension[1].

---

[1] See *Introduction to Electrodynamics* by David Griffiths, 3rd edn, Pearson, London, 2003, for an elaborate discussion of the Laplace equation.

The fact that the potential at a point is the average of the potentials at the neighbouring points implies that local maxima or minima are not possible for the electrostatic potential in a region which satisfies the Laplace equation.

The finite difference method (also known as successive over-relaxation) used to solve the Laplace equation (or any PDE) for given boundary conditions proceeds as follows:

(i) Choose a rectangular grid over the region in which you want to solve the PDE.

(ii) Choose initial values of the potential V for all the grid points. The values at the boundary of the grid are already given.

(iii) Calculate the new values of the potential at each grid point using (10.6).

(iv) Once the calculation of V using (10.6) has been completed for all the grid points, we replace the initially assumed values of V with the newly calculated values.

(v) Steps (iii) and (iv) are repeated many times until the potential at each grid point converges. To check the convergence, we can sum the difference of the V values over the entire grid for two successive iterations.

As an illustration of this method, consider the following boundary-value problem:

The potential at the edges of a square with sides that are 20 cm long is as shown in the figure given below:

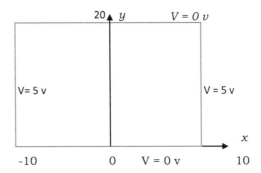

As can be seen in this figure, the potential along the boundaries marked by the lines $x = -10$ and $x = +10$ is 5 V, whereas the potential along the boundaries marked by the lines $y = 0$ and $y = 20$ is 0 V.

The program given below is an implementation of the finite difference method that solves the Laplace equation and determines the potential at the grid points inside the square.

```
'Laplace equation';
'solution using grid method';
clear;
clf;
n=10;
tolerance=0.1;
err=1000;
for i=1:2*n+1
    for j=1:2*n+1
        x(j)=j-(n+1);
        y(i)=i-1;
        v(i,j)=0; 'initializing the grid';
        v1(i,j)=0;
    end
end
for i=1:size(y,2)
    v(i,1)=5;
    v1(i,1)=5;
    v(i,size(x,2))=5; 'specifying the boundary conditions';
    v1(i,size(x,2))=5;
end
'starting successive over relaxation';
while err>tolerance
    err=0;
for i=2:size(y,2)-1
    for j=2:size(x,2)-1
        v1(i,j)=0.25*(v(i-1,j)+v(i+1,j)+v(i,j+1)+v(i,j-1));
        err=err+abs(v(i,j)-v1(i,j));
    end
    end
    v=v1;
    clf;
    mesh(x,y,v1)
    pause(0.5);
    end
```

After a few iterations, the potential at the grid points within the square is shown as a function of the x and y coordinates in figure 10.3.

The final solution, i.e. after the potentials at all grid points have converged to stable values, is shown in figure 10.4.

Note that the solution in figure 10.4 does not have any local maxima or minima, as expected, and is in the shape of a saddle. MATLAB has a provision for rotating the axes and viewing the plot from a different angle. This rotation can be performed

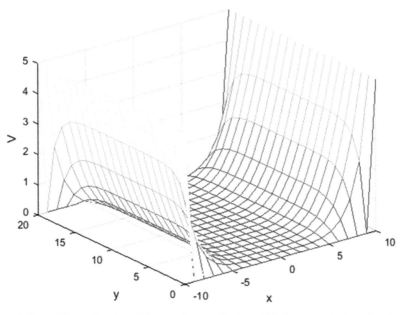

Figure 10.3. Potential as a function of the x and y coordinates within the square (after a few iterations).

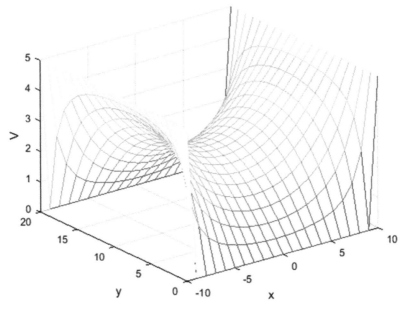

Figure 10.4. The solution to the given boundary-value problem.

by clicking 'Rotate 3D' under the 'Tools' tab in the MATLAB figure window and clicking and dragging inside the figure window. Figure 10.5 shows the solution from a different angle. Viewing a surface from at least two different angles leads to a better understanding of the solution.

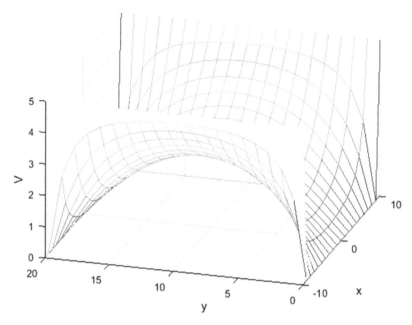

Figure 10.5. A view of figure 10.4 from a different angle.

The boundary-value problem mentioned above has an analytical solution[2] given by:

$$V(x, y) = \frac{4V_o}{\pi} \sum_{n=1,3,5..} \frac{\cosh\left(\dfrac{n\pi x}{a}\right)}{\cosh\left(\dfrac{n\pi b}{a}\right)} \sin\left(\dfrac{n\pi y}{a}\right). \tag{10.7}$$

Here, V_o is the potential at the left and right edges ($V_o = 5$ v in the program) and a and b are the dimensions along the x- and y-axes, respectively. Note that even though an analytical solution exists for this problem (10.7), it is not very illuminating. Equation (10.7) is an infinite series and the numerical value has to be evaluated by truncating the series after a finite number of terms. In many such problems, we find that there is no practical difference between an 'analytical' and a 'numerical' solution.

10.4 A finite difference method for PDEs involving both spatial and temporal derivatives

In the previous section, we solved a partial differential equation using the finite difference method. However, in the Laplace equation, as well as the time-independent Schrödinger equation, we only have derivatives with respect to

---

[2] See chapter 3 of *Introduction to Electrodynamics* by David Griffiths, 3rd edn, Pearson, London, 2003, for details.

spatial coordinates and we can choose the step size in each direction independently, even though it may be convenient to choose the step size such that it has the same value for both the coordinate axes. A complication arises with respect to the convergence of the solution if a partial differential equation has derivatives with respect to both time and space. For instance, consider the partial differential equation given below:

$$\frac{\partial u}{\partial t} = c \frac{\partial u}{\partial x} \tag{10.8}$$

subject to the condition $u(x,0) = u_o(x)$, i.e. we are given the form of the function u at time $t = 0$, and want to determine $u(x)$ at a later time, t. Here, the parameter c has the dimensions of velocity.

By replacing the derivatives in (10.8) with finite differences, we obtain:

$$\frac{u(x, t + \delta t) - u(x, t)}{\delta t} = c \frac{[u(x + \delta x, t) - u(x, t)]}{\delta x}. \tag{10.9}$$

Equation (10.9) can be rewritten to obtain $u(x, t+\delta t)$:

$$u(x, t + \delta t) = u(x, t) + c\delta t \left[\frac{u(x + \delta x, t) - u(x, t)}{\delta x} \right]. \tag{10.10}$$

Equation (10.10) gives us the form of u at a time $t+dt$, in terms of its form at time t. Let us assume that

$$u(x, 0) = \exp(ikx). \tag{10.11}$$

This is not an unreasonable assumption. Any well-behaved function can be represented in terms of its Fourier components; (10.11) is a Fourier component. Since we are dealing with a linear differential equation, any linear combination of $\exp(ikx)$ that will generate any function will also satisfy the PDE. Any function can be represented as a linear combination of different Fourier components.

By substituting (10.11) into (10.10) and taking $t = 0$, we obtain:

$$u(x, \delta t) = \exp(ikx) + c\delta t \left[\frac{\exp(ikx + ik\delta x) - \exp(ikx)}{\delta x} \right]. \tag{10.12}$$

Equation (10.12) can be rewritten as:

$$u(x, \delta t) = \exp(ikx) \left[1 + \frac{c\delta t}{\delta x} (\exp(ik\delta x) - 1) \right] = G(k)\exp(ikx). \tag{10.13}$$

$G(k)$ is a growth factor. After n iterations of (10.13), the solution has grown by a factor G^n. If $|G(k)| > 1$, our solution has 'blown up', i.e. it is unstable. We have to choose a step size for x and t that yields stable solutions. Since k, the wave vector, is a variable, we need to consider a worst-case scenario in which the stability condition $|G(k)| \leqslant 1$ is likely to be violated. On inspecting (10.13), we find that the worst-case scenario corresponds to $k\delta x = \pi$. Substituting this into (10.13), we get:

$$G(k) = 1 - \frac{2c\delta t}{\delta x}. \qquad (10.14)$$

For a stable solution, $|G(k)|$ should be less than one, which gives us the condition:

$$\frac{c\delta t}{\delta x} \leqslant 1. \qquad (10.15)$$

Equation (10.15) tells us that we cannot choose δx and δt independently, but must choose values such that (10.15) is satisfied. It turns out that the same condition should be applied to all partial differential equations (e.g. the wave equation and the time-dependent Schrödinger equation) that involve both spatial and temporal derivatives.

Let us now derive the stability condition for the wave equation and show that we end up with the same condition. The one-dimensional wave equation is given by:

$$\frac{\partial^2 y}{\partial t^2} = c^2 \frac{\partial^2 y}{\partial x^2}. \qquad (10.16)$$

By replacing the derivatives by the corresponding finite differences, we obtain:

$$\frac{y(x, t + \delta t) + y(x, t - \delta t) - 2y(x, t)}{\delta t^2}$$
$$= c^2 \left[\frac{y(x + \delta x, t) + y(x - \delta x, t) - 2y(x, t)}{\delta x^2} \right]. \qquad (10.17)$$

We can rewrite (10.17) to obtain an expression for $y(x,t+\delta t)$ in terms of the displacement y at earlier times.

$$y(x, t + \delta t) = -y(x, t - \delta t) + 2y(x, t)$$
$$+ c^2 \delta t^2 \left[\frac{y(x + \delta x, t) + y(x - \delta x, t) - 2y(x, t)}{\delta x^2} \right]. \qquad (10.18)$$

Equation (10.18) can be further simplified to:

$$y(x, t + \delta t) = -y(x, t - \delta t) + 2y(x, t)\left[1 - \frac{c^2 \delta t^2}{\delta x^2} \right]$$
$$+ \frac{c^2 \delta t^2}{\delta x^2}[y(x + \delta x, t) + y(x - \delta x, t)]. \qquad (10.19)$$

Just as we did for (10.8), we will consider only one Fourier component of the solution. Since solutions to the wave equation have the form $f(kx - \omega t)$, we consider the solution to be of the form:

$$y(x, t + \delta t) = \exp[i(kx - \omega t)]. \tag{10.20}$$

By substituting (10.20) into (10.19), we obtain:

$$y(x, t + \delta t) = -e^{i(kx-\omega t)}e^{i\omega t} + 2e^{i(kx-\omega t)}\left[1 - \frac{c^2\delta t^2}{\delta x^2}\right]$$
$$+ \frac{c^2\delta t^2}{\delta x^2}e^{i(kx-\omega t)}[e^{ik\delta x} + e^{-ik\delta x}]. \tag{10.21}$$

Following (10.13), we can rewrite (10.21) as:

$$y(x, t + \delta t) = G(k, \omega)e^{i(kx-\omega t)}, \tag{10.22}$$

where:

$$G(k, \omega) = -e^{i\omega t} + 2\left[1 - \frac{c^2\delta t^2}{\delta x^2}\right] + \frac{c^2\delta t^2}{\delta x^2}[2\cos(k\delta x)]. \tag{10.23}$$

To obtain a stable solution, we require $G(k, \omega) \leqslant 1$. Since both k and ω are variables, we have to consider the 'worst-case scenario', in which the stability condition is likely to be violated. The worst-case scenario occurs for $\cos(k\delta x) = -1$ and $\exp(i\omega t) = -1$.

By substituting these values into (10.23), we obtain:

$$G(k, \omega) = 1 + 2\left[1 - \frac{c^2\delta t^2}{\delta x^2}\right] - 2\frac{c^2\delta t^2}{\delta x^2} = 3 - 4\frac{c^2\delta t^2}{\delta x^2}. \tag{10.24}$$

Again, we see that the condition to be satisfied for stability is:

$$\left(\frac{c\delta t}{\delta x}\right)^2 \leqslant 1, \tag{10.25}$$

which is same as the stability condition (10.15) we got earlier.

The only complication with iterating (10.18) to determine the solution to the wave equation is that we need to know the displacement at times t and $t - \delta t$. The initial conditions for the wave equation are usually specified in terms of $y(x,0)$ and the initial velocity at all points on the string, $\frac{\partial y}{\partial t}|_{t=0}$. We can write the initial velocity in terms of finite differences and hence eliminate $y(x, t-\delta t)$, which occurs in (10.18).

The following MATLAB program is an implementation of the finite difference method for solving the wave equation:

```matlab
'program to simulate wave motion';
clear all;
maxtime=0.25; % maximum simulation time;
dt =0.0004; % time increment;
L=1; % string length;
dx=0.01; % space increment;
x=0:dx:L;
T=10; % tension;
mu=0.1; %mass density;
c=(T/mu)^0.5;    % wave velocity;
epsilon=(dt*c/dx)^2;  % needs to remain less than 1 for stability;
time=0.0;
u(1)=0;
u(size(x,2))=0;
% following line is to specify the initial shape of the string;
for i=2:size(x,2)-1
    u(i)=0.1*exp(-100*(x(i)-0.5)^2);
end
% following line is to specify the initial velocity of the string at
%each point
for i=1:size(x,2)
    du(i)=0.0;
end
j=1;
vidObj = VideoWriter('wave.avi');
    open(vidObj);
% The above command is to open a new avi file. The results of the
%simulation will be stored in that file.
while time<maxtime
    time=time+dt;
    unew(1)=0.0;
    unew(size(x,2))=0.0;
    for i=2:size(x,2)-1
        unew(i)=epsilon*(u(i+1)+u(i-1))+2*(1-epsilon)*u(i)-
u(i)+du(i);
    end
    for i=2:size(u,2)
        du(i)=unew(i)-u(i);
    end
    u=unew;
    clf;
    plot(x,u)
axis([0 1 -0.1 0.1])
    pause(0.1);

    currFrame = getframe;
        writeVideo(vidObj,currFrame);
% storing the results in the avi file.
end
```

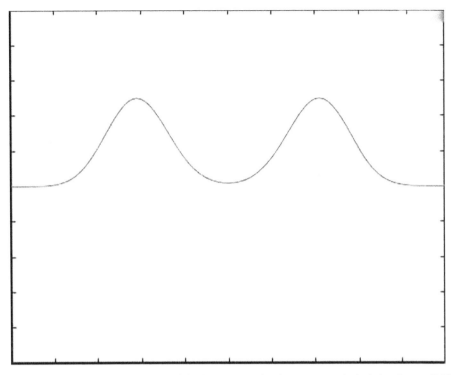

Figure 10.6. Visualization of the output of the 'program to simulate wave motion'. Animation available at https://iopscience.iop.org/book/978-0-7503-3791-5.

You can vary the step size for time and x and watch the simulation in the figure window.

The results obtained by the above program are displayed in figure 10.6.

In the above program, we have assumed that the initial shape of the string is Gaussian. You will have noticed that the single Gaussian peak splits into two and the two pulses travel in opposite directions. This is because the wave equation permits two solutions that correspond to waves travelling in opposite directions. Note also that the pulse undergoes a phase change of π when it is reflected at the end of the string.

Exercises

10.1. Approximate the derivatives in the equation $2\frac{\partial^2 u}{\partial x^2} + \frac{\partial^2 u}{\partial y^2} - \frac{\partial u}{\partial x} = 2$ by finite central differences and obtain the equation which can be used to solve this equation by successive over-relaxation.

10.2. Write the partial differential equation $\nabla^2 u + \lambda u = 0$ as a finite difference equation. Here, λ is a constant and ∇^2 is the Laplacian in two dimensions.

IOP Publishing

Computational Methods Using MATLAB®
An introduction for physicists
P K Thiruvikraman

Chapter 11

Nonlinear dynamics, chaos, and fractals

11.1 History of chaos

As pointed out in chapter 9, the equations of physics, e.g. the wave equation or the Schrödinger equation, seem to be linear. However, note that the wave equation is derived by assuming that the amplitude of the wave is negligible when compared with its wavelength. Similarly, the equation of motion for the simple pendulum is linearized by assuming that the amplitude of oscillation is small. Traditionally, physicists and mathematicians have concentrated on solving linear equations, as such equations permit closed-form or analytical solutions and also obey the principle of superposition (all of wave optics is based on the application of the principle of superposition). However, nature is inherently nonlinear, as seen in the examples mentioned above (the simple pendulum and the wave equation).

For the physicists and mathematicians of the eighteenth and nineteenth centuries, Newtonian mechanics seemed to describe a 'clockwork' Universe. In fact, the great French mathematician, Pierre Simon Laplace, in his book *A Philosophical Essay on Probabilities*, published in 1815, said,

> *'We may regard the present state of the universe as the effect of its past and the cause of its future. An intellect which at a certain moment would know all forces that set nature in motion, and all positions of all items of which nature is composed, if this intellect were also vast enough to submit these data to analysis, it would embrace in a single formula the movements of the greatest bodies of the universe and those of the tiniest atom; for such an intellect nothing would be uncertain and the future just like the past would be present before its eyes.'*

This 'intellect' has since been named 'Laplace's demon'. This demon was only exorcised in the 20th century. Heisenberg struck the first blow with his uncertainty principle and the demon reeled from it. According to Heisenberg's uncertainty principle, it is impossible to make a simultaneous measurement of both the position

doi:10.1088/978-0-7503-3791-5ch11

and the corresponding momentum of a particle with arbitrary precision as δx and δp_x (the uncertainties in the x coordinate and the x component of the momentum, respectively) meet the condition $\delta x \delta p_x \geq \hbar/2$. If we do not know the initial position and momentum of the particle precisely, then obviously this uncertainty will affect our knowledge of its subsequent motion.

Furthermore, it was first shown by Edward Lorenz in the 1960s that a slight difference in the initial conditions of a nonlinear system can lead to a large difference in the state of the system after some time. This sensitivity to initial conditions is also known as the 'butterfly effect'[1]. This sensitivity to initial conditions is the hallmark of chaos, as we can no longer predict the long-term behaviour of such a system. The weather is a familiar example of such a chaotic system.

What is the fundamental requirement for a system to be chaotic? This is a question that arises naturally in the study of nonlinear dynamics. The requirement can be stated in terms of a phase-space plot of the state of the system. In the phase-space portrait of a one-dimensional system, we plot the position along the x-axis and the corresponding momentum along the y-axis. For example, the phase-space trajectory of an undamped simple harmonic oscillator is an ellipse (figure 11.1).

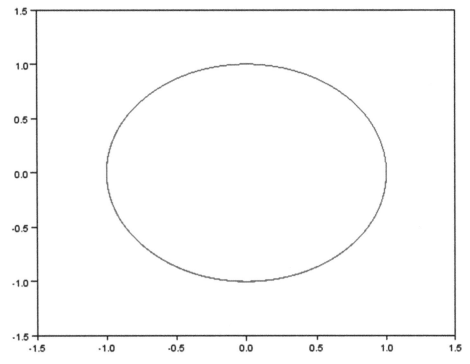

Figure 11.1. Phase-space portrait of a simple harmonic oscillator.

---

[1] *Chaos: The Making of a New Science* by James Gleick, Viking Books, New York, 1987, is an excellent popular account of nonlinear dynamics and chaos. Another well-written book on the same topic is *Does God Play Dice* by Ian Stewart, 2nd edn, Penguin Books, London, 1997.

This is because the energy of an undamped simple harmonic oscillator is a constant and is given by:

$$E = \frac{kx^2}{2} + \frac{p^2}{2m}.\tag{11.1}$$

This is the equation for an ellipse with semi-major and semi-minor axes of $\sqrt{\frac{2E}{k}}$ and $\sqrt{2mE}$, respectively. A closed trajectory such as an ellipse signifies that the system exhibits periodicity and is obviously not chaotic.

If the harmonic oscillator is subject to a damping force, its trajectory in phase space is a spiral, as can be seen in figure 11.2.

Chaos is not possible if the phase space has only two dimensions. This is because the trajectory of a system cannot intersect itself in phase space. An intersection would imply that a system that has a repetition of the same position and momentum can choose from among multiple paths. However, according to Newton's laws, the trajectory of a system is unique for a given combination of initial position and momentum. Therefore, chaos is only possible if the phase space has more than two dimensions. This is possible for systems whose motion occurs in two or three dimensions. The corresponding dimensionalities of the phase spaces would be four and six, respectively. In such cases, the orbit in phase space can be confined within a given volume (conservative system) without being periodic. The trajectory can then come arbitrarily close to its position at an earlier time and then 'escape' into a different dimension.

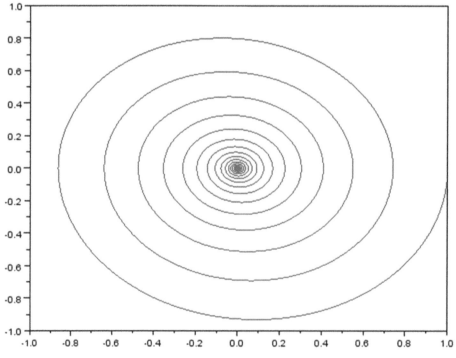

Figure 11.2. Phase-space portrait of a damped oscillator.

Historically, one of the first systems in which chaos was noticed was a system of coupled nonlinear differential equations used to model the weather. Edward Lorenz was trying to model weather using the system of coupled nonlinear differential equations given below:

$$\frac{dx}{dt} = \sigma(y - x) \tag{11.2}$$

$$\frac{dy}{dt} = px - y - xz \tag{11.3}$$

$$\frac{dz}{dt} = -\beta z + xy. \tag{11.4}$$

Edward Lorenz used these equations to model the weather in the 1960s using the parameters:

$$\sigma = 10, \quad \beta = 8/3, \quad \rho = 28$$

It should be noted that the same set of equations does not exhibit chaos if the values of the three parameters σ, β, and ρ, are in some other range.

The following program gives a numerical solution for the Lorenz equations using the simple Euler method:

```
'lorenz equations';
clf; 'clears the figure window'
clear all;

x(1)=0;
y(1)=1;
z(1)=0;
dt=0.001;
for i=2:100000,
    xdot=10*(y(i-1)-x(i-1));
    ydot=28*x(i-1)-y(i-1)-x(i-1)*z(i-1);
    zdot=-(8/3)*z(i-1)+x(i-1)*y(i-1);
    x(i)=x(i-1)+xdot*dt;
    y(i)=y(i-1)+ydot*dt;
    z(i)=z(i-1)+zdot*dt;
end

plot3(x,y,x);
```

Figure 11.3 shows Lorenz's famous 'butterfly' diagram. It was perhaps this figure which led him to coin the term 'butterfly effect', which signifies the sensitive dependence on the initial conditions of chaotic systems. This figure also illustrates that if the state space of the system is three-dimensional, complex behavior is

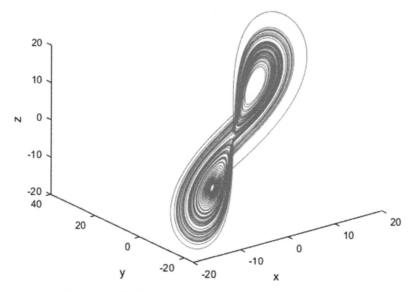

Figure 11.3. The famous butterfly diagram discovered by Lorenz.

possible, even if the system is constrained within some region of the state space—since it has a third dimension to escape into.

For chaos to be exhibited, a minimum of three conditions have to be met:

(i) The phase space for the system should have more than two dimensions.
(ii) The equations governing the system should be nonlinear.
(iii) The system should be in a region of parameter space which exhibits chaos.

The first two conditions can be easily verified for a given system. However, even if the number of dimensions is four (the simplest example of a four-dimensional phase space is a double pendulum), and the equations governing the system are nonlinear, the system should be placed in a suitable region of parameter space for it to exhibit chaos. In the case of the double pendulum, we need to make the oscillatory amplitude of a double pendulum large for it to exhibit chaos. For other systems, the conditions required may not be that obvious. In general, we can state that chaos occurs wherever a nonlinear term dominates in the equations of motion.

An easier way to study chaos is through the study of one-dimensional maps.

11.2 The logistic map

We will begin our study of chaos by looking at a simple equation known as a logistic map. The logistic map arose from attempts to model populations (of insects, etc). The logistic map is given by:

$$p_{n+1} = p_n(a - bp_n). \tag{11.5}$$

This equation relates the population of an insect colony at particular time (the $n+1$th generation) to the population at an earlier time (the nth generation). Note that time moves in discrete steps in this model. The insect population can also be modelled by writing down a differential equation (this would correspond to a continuous variation in time, but the continuous variation in population is not realistic!).

The first term models the growth, while the second term is related to the decrease in the population due to overcrowding or competition. Analysis of the behaviour of this map is easier if we change the variables.

Let

$$p_n = \frac{a}{b} x_n \tag{11.6}$$

$$r = \frac{a}{4}. \tag{11.7}$$

By substituting (11.6) and (11.7) into (11.5), we obtain:

$$x_{n+1} = 4r x_n (1 - x_n). \tag{11.8}$$

Here, x and r are restricted to the interval $[0,1]$ to avoid negative values of x (which would correspond to negative values of the population).

This map was studied extensively in the late 1970s by Feigenbaum, who discovered that even a seemingly trivial map can produce interesting results. To study (11.8) numerically, we choose a value for r and an initial value for x: let $r = 0.25$ and $x_0 = 0.5$. Substituting these values into (11.8) gives the value of x_1 (the left-hand side), which is 0.25. Substituting x_1 into the equation yields 0.1875. In fact, the value of x decreases monotonically in subsequent iterations and tends to zero after some iterations. Figure 11.4 shows a plot of x versus the number of iterations.

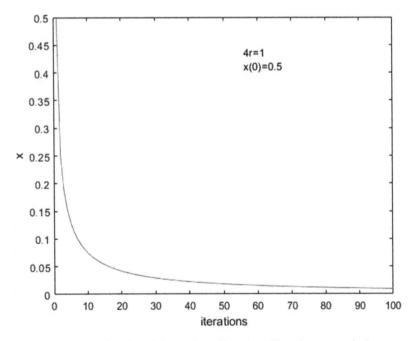

Figure 11.4. x as a function of the number of iterations. Note the monotonic decrease.

We now repeat our numerical study of the map by changing r to 0.5, but we start with $x_0 = 0.5$ as before. Substituting the values of x and r in (11.8) gives us 0.5 as the output. The output value will then obviously be equal to 0.5 for all subsequent iterations. Figure 11.5 gives a plot of this seemingly uninteresting result.

If we start with a different initial condition, the values converge to the fixed point after a few iterations, as seen in the 'cobweb' diagram (figure 11.6).

We further increase r to 0.625 and notice that something different is happening now. The value of x increases from its initial value of x and oscillates before settling at a final value of 0.6 (figure 11.7).

In all the cases considered above, the main commonality is that there is a 'fixed point' to which the value of x converges after many iterations. The fixed point was zero for low values of r, but for higher values of r, the fixed point is nonzero and seems to depend on r. The fixed point for (11.8) can be calculated analytically. For a fixed point, we have $x_{n+1} = x_n$:

$$x_{n+1} = 4rx_n(1 - x_n) = x_n. \tag{11.9}$$

Equation (11.9) is a quadratic equation in x with two roots. The two roots are $x = 0$ and

$$x = 1 - \frac{1}{4r}. \tag{11.10}$$

Recall the discussion about the iterative fixed-point method in chapter 3. We see that the condition for convergence (section 3.3) is $f'(x) < 1$. By applying the condition for convergence to (11.9), we obtain the condition:

$$f'(x) = 4r - 8rx < 1. \tag{11.11}$$

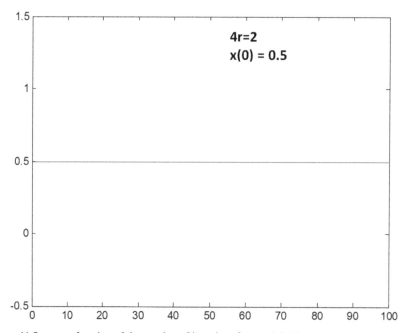

Figure 11.5. x as a function of the number of iterations for $r = 0.5$. Note the constant value of x.

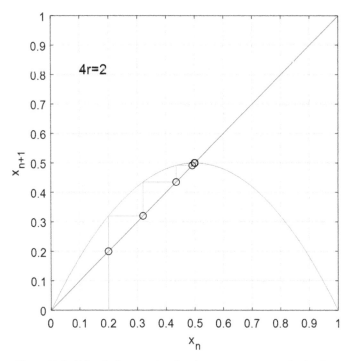

Figure 11.6. Cobweb diagram showing the convergence to a fixed point.

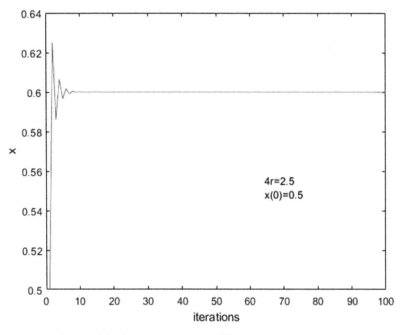

Figure 11.7. Note that the value oscillates about the fixed point.

This means that the fixed point at $x = 0$ is stable (convergent) if $x < 0.25$. The other root is stable if

$$f'(x) = 4r - 8r\left(1 - \frac{1}{4r}\right) < 1 \text{ or } 4r > 1 \tag{11.12}$$

This is the reason the values converge to the nonzero root for $r > 0.25$.

If we increase r to a value greater than 0.625, the oscillations die out after a larger number of iterations, as can be seen in figures 11.8 and 11.9.

If we increase further to $4r = 3.25$, the oscillations become the primary feature (figure 11.10).

Another way of looking at this is to say that $x_{n+2} = x_n$. By imposing this condition on (11.8), we obtain:

$$x_{n+2} = 4rx_{n+1}(1 - x_{n+1}) = 4rx_n(1 - x_n)(1 - 4rx_n(1 - x_n)). \tag{11.13}$$

Equation (11.13) is a quartic equation which has four roots. Two of these roots are $x = 0$ and $x = 1 - \frac{1}{4r}$.

After factoring out these two roots, we find that the remaining two roots are:

$$\frac{4r + 1 \pm \sqrt{(4r - 3)(4r + 1)}}{8r}. \tag{11.14}$$

Therefore, the oscillation (or two-cycle as it is called in the literature) is only possible for $4r > 3$. This phenomenon is also known as period doubling. A further increase of the r value leads to a further doubling, which gives rise to a four-cycle (figure 11.11).

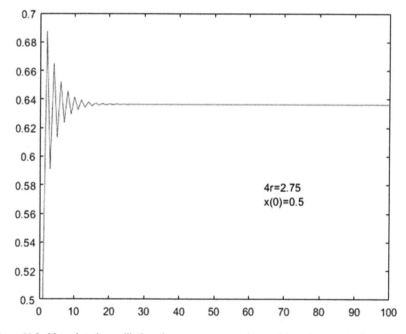

Figure 11.8. Note that the oscillations become more prominent with an increase in the value of r.

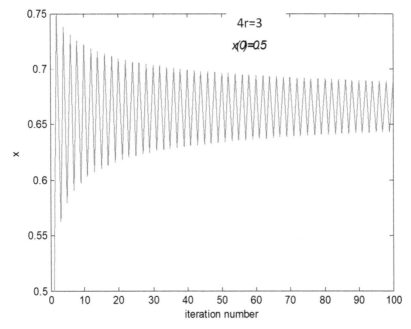

Figure 11.9. Note that the oscillations do not die down for $4r = 3$.

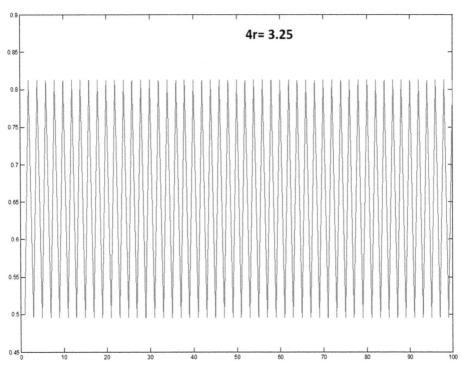

Figure 11.10. Oscillations about the fixed point.

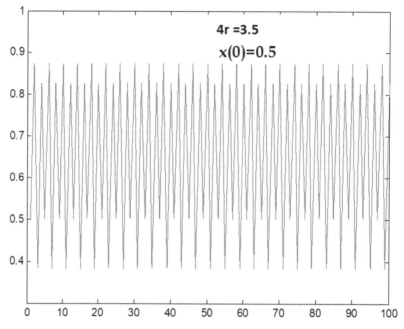

Figure 11.11. Logistic map that exhibits a period of four iterations.

A further increase of r gives rise to a more rapid period doubling, and the system soon becomes chaotic. Figure 11.12 shows the values of x for $4r = 3.75$.

Figure 11.12 seems to exhibit chaotic behaviour, as we cannot identify any periodicity in it. We can confirm the chaotic behaviour by checking the values obtained for two initial conditions which differ very marginally. Figure 11.13 shows the outputs obtained for two slightly different initial values (0.6 and 0.599). Note that the two paths are vastly different after a few iterations.

Each doubling of the period is known as a 'bifurcation'[2]. Figure 11.14 shows a bifurcation diagram for the logistic map which includes all of the various bifurcation points.

The initial bifurcation (or period doubling) happens for $4r = 3$ and the second doubling happens at around $4r = 3.49$, but the subsequent doublings happen very rapidly and the bluish region means that we have n-cycles, where n is very large. We also notice some white 'windows' in the bluish region. For instance, at around $4r = 3.75$, we have a three-cycle. The onset of chaos happens beyond this point. A fascinating thing about the logistic map is that a completely deterministic map gives rise to chaos! We would have no way of predicting this chaotic behavior from the apparently trivial nature of the logistic map. The only hint of chaos lies in the nonlinearity of the map.

A physical example of this period doubling is a leaking faucet. We are all familiar with the example of a faucet which is not completely closed and drips periodically. Initially, the drops are very periodic. If we carefully open the tap very gently, then

---

[2] The complete theory of chaotic systems is discussed extensively in *Nonlinear Dynamics and Chaos*, Steven Strogatz, Westview, Boulder, CO, 2000.

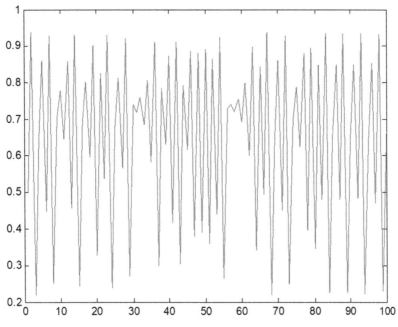

Figure 11.12. Successive values of x for $4r = 3.75$.

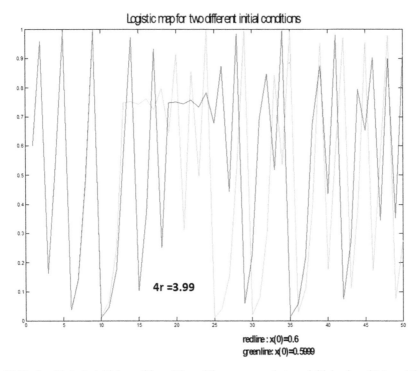

Figure 11.13. Sensitivity to initial conditions. The red line corresponds to an initial value of 0.6, and the green line corresponds to an initial value of 0.5999.

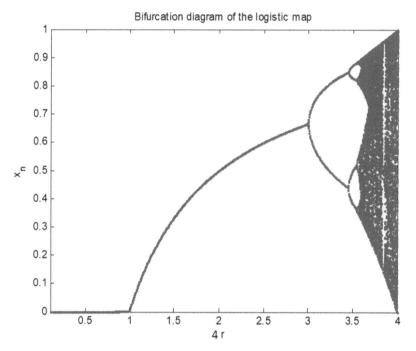

Figure 11.14. Bifurcation diagram for the logistic map.

we will have a period doubling at some point, and subsequent period doublings can also be observed[3].

Figure 11.13 shows the sensitive dependence on the initial conditions of a chaotic system. Can we get a quantitative measure of the divergence of the trajectories and hence obtain a way to diagnose chaos?

11.3 The Lyapunov exponent

The Lyapunov exponent can be used to characterize the divergence of two nearby trajectories. If Δx_0 is the initial difference between two values of x and Δx_n, the difference after n iterations of a map, then an exponential divergence would imply:

$$|\Delta x_n| = |\Delta x_0|\exp{(\lambda n)}. \qquad (11.15)$$

However, direct application of this equation is difficult, because:
1. The rate of separation of the trajectories might depend on the choice of x_o.
2. The separation ceases to increase when n is sufficiently large (x has to be within the unit interval).

We can avoid these problems by taking the logarithm of both sides and rewriting (11.15) as:

---

[3] Many such physical examples of chaos can be found in the very concise book, *Deterministic Chaos* by N Kumar, Universities Press, Hyderabad, 1996.

$$\lambda = \frac{1}{n} \ln \left[\frac{|\Delta x_n|}{|\Delta x_0|} \right]. \tag{11.16}$$

We also rewrite the ratio as:

$$\frac{\Delta x_n}{\Delta x_0} = \frac{\Delta x_n}{\Delta x_{n-1}} \frac{\Delta x_{n-1}}{\Delta x_{n-2}} \cdots \frac{\Delta x_1}{\Delta x_0}. \tag{11.17}$$

By substituting (11.17) into (11.16), we obtain:

$$\lambda = \frac{1}{n} \sum_{i=0}^{n-1} \ln \left| \frac{\Delta x_{i+1}}{\Delta x_i} \right|. \tag{11.18}$$

The ratio of the Δx values can be written as:

$$\frac{\Delta x_{i+1}}{\Delta x_i} = \frac{dx_{i+1}}{dx_i} = f'(x_i). \tag{11.19}$$

In the case of the logistic map,

$$f'(x_i) = 4r(1 - 2x_i). \tag{11.20}$$

By substituting (11.20) into (11.19), we obtain:

$$\lambda = \frac{1}{n} \sum_{i=0}^{n-1} \ln|4r(1-2x_i)|. \tag{11.21}$$

The following program calculates the Lyapunov exponent for various values of r and plots it as a function of r.

```
'Lyapunov exponent for logistic map';
clear;
x(1)=0.6;
k=1;
r=0.7;

while r<1

for i=1:1000000
    x(i+1)=4*r*x(i)*(1-x(i));
end
n=size(x,2);
lambda(k)=0;
for i=1000:n-1
    m=abs(1-2*x(i));
    lambda(k)=lambda(k)+(1/n)*(log(4*r*m));
end
r1(k)=r;
k=k+1;
r=r+0.001;
end
plot(r1,lambda)
```

11-14

The plot generated by running the program gives the output shown in figure 11.15.

In the above program, we ignored the first 1000 values in order to make the value of the exponent independent of the initial value of x. You can experiment with the initial value and the value of n and see whether there are any variations in the results.

From the graph, it can be seen that the Lyapunov exponent only becomes positive for $r > 0.893$, indicating the onset of chaos. However, even in the range from 0.893 to one, there are windows of stability. As r tends to one, the Lyapunov exponent becomes larger, indicating that the system is becoming more and more chaotic. It is fascinating that such a rich chaotic structure is buried in such a simple equation!

The logistic map is just one example of a one-dimensional map; many other maps have been extensively studied by mathematicians.

Example 11.1

Consider the one-dimensional map $x_{n+1} = \tan x_n$.

(a) Determine the approximate value of one fixed point of this map apart from the fixed point at $x = 0$.

(b) Determine the stability of the fixed point you calculated.

(c) Determine the stability of the fixed point at the origin.

Answer: (a) The other fixed point will occur when $x = \tan(x)$. A graph of $\tan(x)$ versus x will help us to determine this fixed point (figure 11.16).

It is clear from the graph that the fixed point is between π and $3\pi/2$; therefore, we use bisection with initial values of π and $3\pi/2$. After one iteration, we find that the

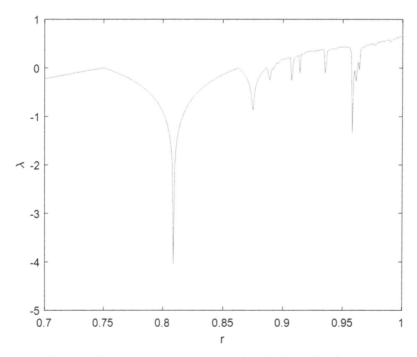

Figure 11.15. Lyapunov exponent as a function of r for the logistic map.

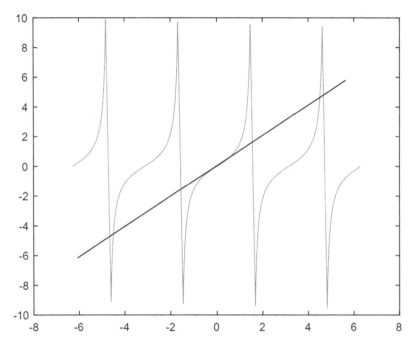

Figure 11.16. Solution for the equation $x = \tan(x)$.

fixed point is bounded by $5\pi/4$ and $3\pi/2$. One more iteration indicates that the fixed point is between $11\pi/4$ and $3\pi/2$. A few more iterations indicate that the fixed point is close to 1.43π.

(b) Stability can be obtained by evaluating $f'(x)=\sec^2(x) > 1$. A cobweb diagram also shows that it is unstable.

(c) At the origin, $\sec^2(x) = 1$; therefore, we have to use the cobweb diagram to determine the stability (figure 11.17). The cobweb diagram seems to indicate that the value of x returns to a value close to zero after many iterations. In that sense, it is stable.

We discuss two more maps in order to highlight the perils hidden in the study of maps. Consider the map $x_{n+1} = 2x_n$ mod 1. This is the map we discussed in example 11.1. Just to recall the result obtained then:

If we choose 0.6 as the initial number, we get 1.2 after multiplying by two and the decimal part is 0.2. On repeated multiplication by two, we get: 0.4,0.8,0.6,0.2... For the first 40 iterations or so, the pattern repeats, but at some point, you start seeing numbers such as 0.6001, 0.2002, 0.4004 ... and after some time, the number simply becomes zero.

The problem lies in the finite precision with which numbers are stored on a computer. Since this map shifts the bits to the left after each iteration and replaces the last bit by zero, the solution is to choose the rightmost bit randomly after each iteration, so that it is zero or one (this is equivalent to specifying the initial number to infinite precision).

The same problem occurs in the case of the tent map defined by:
$f(x) = 2x$ *for* $0 \leqslant x \leqslant \frac{1}{2}$
$f(x) = 2(1-x)$ *for* $\frac{1}{2} \leqslant x \leqslant 1$.

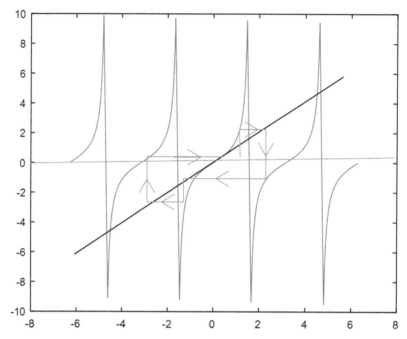

Figure 11.17. Cobweb diagram for example 11.1(c).

The more general version of the tent map replaces the factor of two with r (where $r < 2$). Figure 11.18 shows the output of the tent map after each iteration.

In figure 11.18, the initial value of x was chosen to be 0.43423. The problem here is that if the value of x is equal to 0.5 at some stage, it will become equal to zero in the next iteration and remain at zero forever. The root cause of this problem is that 0.43423 cannot be represented exactly in binary (recall the discussion in section 1.2). Therefore, even though the computer displays 0.434230000000000 (if we use the 'long' format), internally, some of the least significant bits are nonzero. These are shifted to the right when the number is multiplied by two. This problem is obvious if you print out the number at each iteration. For instance, the 7th value of x is 0.209280000000000, but the subsequent value is 0.418559999999999 (instead of 0.41856).

This shows that we have to be extremely cautious in interpreting the results printed out by computers. At this stage, you can be excused if you have a nagging doubt: is the chaotic behaviour of many systems (as seen in numerical simulations) an intrinsic quality of these systems, or is it due to the finite precision of computers? If it were the latter, you might as well scrap all the computers in the world (or at least, stop using them to study chaotic systems). Fortunately, this is not the case[4].

---

[4] See 'Chaos in the Lorenz equations: a computer assisted proof,' K Mischaikow K and M Mrozek, *Bull. Am. Math. Soc. (N.S.)* **33**, 1995, 66–72 for details.

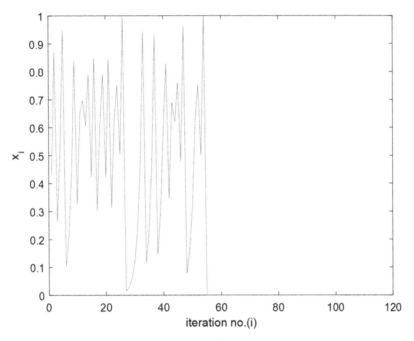

Figure 11.18. Output for the tent map.

11.4 Differential equations: fixed points

The study of differential equations (including systems of equations) throws up fixed points, and the stability of those fixed points is an important consideration, keeping in mind systems of equations such as those given by Lorenz.

Generally, first-order equations are extensively studied in the field of chaos, as even a second-order differential equation can be reduced to a set of two first-order differential equations.

The first-order equations are of the form

$$\frac{dx}{dt} = f(x). \tag{11.22}$$

A second-order differential equation such as Newton's second law $\frac{d^2x}{dt^2} = \frac{F(x)}{m}$, can be written as two first-order differential equations:

$$\frac{dv}{dt} = \frac{F(x)}{m} \text{ and } \frac{dx}{dt} = v. \tag{11.23}$$

A second-order differential equation such as the equation for the forced harmonic oscillator:

$$\frac{d^2x}{dt^2} + \gamma\frac{dx}{dt} + \omega_0^2 x = \frac{F_o \cos \omega t}{m} \tag{11.24}$$

can be written as the following system of three equations:

$$\frac{dx_2}{dt} = \frac{F_o \cos x_3}{m} - \gamma x_2 - \omega_0^2 x_1 \tag{11.25}$$

$$\frac{dx_1}{dt} = x_2 \tag{11.26}$$

$$x_3 = \omega t. \tag{11.27}$$

Hence, a system of first-order differential equations can represent any higher-order differential equation; this formalism is general enough to merit a detailed discussion.

The fixed point of equation (11.22) is the value of x for which $f(x) = 0$.

The time evolution of the system represented by equations such as (11.23) can be represented by a phase-space diagram that represents the quantity x along the horizontal axis and the temporal derivative of x along the vertical axis.

For instance, consider the equation

$$\dot{x} = 1 - 2\cos x. \tag{11.28}$$

The phase-space portrait for equation (11.28) is shown in figure 11.19. The fixed points of this system are: $\cos(x) = \frac{1}{2}$; i.e. $x = \pm\pi/3$.

Note that the fixed points at $x = \pm 5\pi/3$ are identical to those at $x = \pm\pi/3$. It is meaningful to look at just one interval of 2π (say $-\pi$ to $+\pi$).

The arrows on the horizontal axis indicate the direction in which the system evolves. If dx/dt is positive, x increases, whereas, if dx/dt is negative, x decreases. If the arrows point towards a fixed point from both directions, then it is a stable fixed point (indicated by a filled circle in figure 11.19). If the arrows point away from a fixed point, it is an unstable fixed point (shown by an unfilled circle in figure 11.19). If the arrows point towards a fixed point from one direction and away from it in the other direction, then such a fixed point is said to be half-stable (usually shown by a half-filled circle).

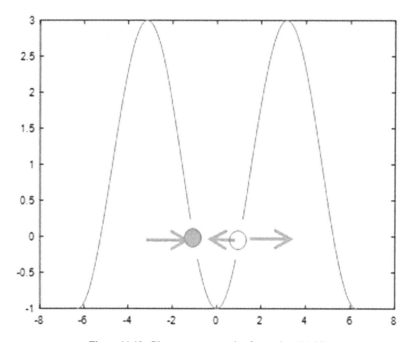

Figure 11.19. Phase-space portrait of equation (11.28).

Differential equations such as (11.28) become more interesting from an analysis perspective if they contain a parameter r. The variation of this parameter produces bifurcations (similar to those seen in the logistic map). We consider bifurcations by considering a few examples.

Example 11.2

Consider the system $\dot{x} = rx - \ln(1 + x)$.

(a) Consider the behavior of the system as r is varied and determine the type of bifurcation which occurs.

(b) At what value of r does the bifurcation[5] occur?

(c) What are the fixed points of the system on either side of the bifurcation point?

Answer: (a) to obtain the fixed points of the given equation, it is convenient to plot rx and $\ln(1+x)$ and look for the points where they are equal (which correspond to fixed points). For $r > 1$, there is only one fixed point, which is at the origin, as shown in figure 11.20.

When r is reduced to less than one, another fixed point can occur (figure 11.21):

(b) The bifurcation occurs at $r = 1$.

(c) For $r > 1$, the only fixed point is at $x = 0$; for $r < 1$, there is one more fixed point, which can be obtained from the equation

$rx = \ln(1+x)$.

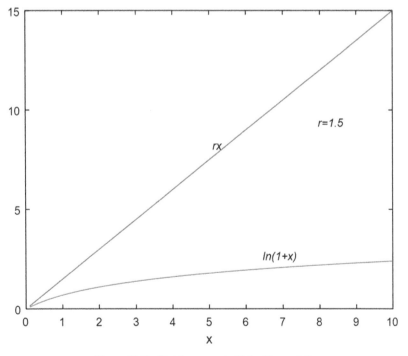

Figure 11.20. Plot for example 11.2 with $r = 1.5$.

---

[5] For a complete discussion of the various types of bifurcation that are possible, please refer to *Nonlinear Dynamics and Chaos*, Steven Strogatz, Westview, Boulder, CO, 2000.

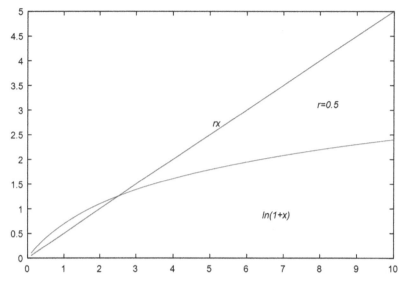

Figure 11.21. Plot for example 11.2 with $r = 0.5$.

For a more detailed and richer study of chaos, we need to look at system of equations (with a minimum of two equations). Such systems of equations have the form:

$$\frac{d\dot{x}_1}{dt} = f(x_1, x_2) \tag{11.29}$$

$$\frac{d\dot{x}_2}{dt} = f(x_1, x_2). \tag{11.30}$$

The behavior of such a system of equations can be studied by plotting vector fields, in which we plot x_1 and x_2 along the two axes and the derivatives are indicated by arrows drawn at every point. In other words, we plot a vector field. Joining the arrows indicates the trajectory at a point.

A detailed study of chaos is essential for a proper understanding of weather and climate change. Climate change has become a topic of great interest, as exemplified by the award of the Nobel Prize in Physics in 2021 for work done in this field.

With the future promising faster computers, we would like to play the part of Laplace's demon, though not in the way Laplace would have intended.

11.5 Fractals

Just like the study of chaos, the study of fractals took off in the second half of the 20th century. The mathematician Benoit Mandelbrot really popularized the study of fractals by writing his book 'The fractal geometry of nature', originally published in French in 1975 under a slightly different title.

Fractals are ubiquitous in nature: snowflakes, trees, and seashells are the most well-known examples. Fractals also make attractive screensavers!

The most striking feature of a fractal is its self-similarity; in other words, the structure of a fractal looks the same at any length scale. For instance, a tree splits into many branches. The branches may split into smaller branches and then twigs. The twigs in turn have leaves and the leaves also exhibit a branching structure.

Figure 11.22. The Cantor set. The black line at the top is 243 pixels long.

Such a self-similar structure can be constructed by the recursive application of a rule. For instance, consider the well-known Cantor set, which is constructed as follows. Draw a straight line of a certain length and (mentally) divide it into three parts. Erase the middle one-third of the straight line. Now consider the remaining two portions and remove the middle one-third in each one of them. Continue this process indefinitely to produce the Cantor set. Figure 11.22 shows the Cantor set.

A MATLAB program was used to generate the Cantor set. The black line at the top of figure 11.22 is 243 pixels long (it makes sense to choose a number that is a power of three, so that the number of pixels to be removed remains an integer until the last iteration). The line below that has the middle one-third removed from the line above, i.e., pixels 82 to 162 are removed (whitened). Each time the black lines are copied to the row below and the middle one-third of each line is removed.

The MATLAB program used to generate the Cantor set is shown below:

```
'cantor set';
k=5;
for i=1:3^k
    for j=1:30
        a(j,i)=1;
    end
end

n=5;
for i=1:3^k
    a(n,i)=0;
end
m=1;

while k>1
    n=n+5;
    k=k-1;
    a(n,:)=a(n-5,:);
for j=1:m
    for i=1+(3*(j-1)+1)*3^k:(3*(j-1)+2)*3^k
        a(n,i)=1;
    end
end
m=m*3;
end
imshow(a)
' imshow displays the matrix a on the figure window';
```

A little bit of explanation is in order. The 'while' loop iterates on the straight line, i.e. we proceed from the initial line to the next stage of the Cantor set. After each stage of the Cantor set, the line is broken into more and more segments. The outer 'for' loop moves from one segment to the next, while the inner 'for' loop moves from one pixel to the next within the segment to be removed. Removing part of the straight line is accomplished by setting the corresponding pixels to one (the pixels that survive are all zero, corresponding to black pixels).

The Sierpinski carpet is a generalization of the Cantor set to two dimensions (figure 11.23).

The Sierpinski carpet is generated using the following steps:

(i) Consider a square of side a (which corresponds to, say, a pixels). This square is then divided into nine equal boxes, and the central box is deleted.

(ii) This process is then repeated for each of the remaining boxes (i.e. each box is divided into nine sub-boxes and the central sub-box is deleted).

(iii) This process (step (ii)) is repeated until the size of the sub-box reaches one pixel.

To implement this in MATLAB and display the output as an image the algorithm given below can be used:

(i) Initialize all the elements of an array (let us call it A) of size $a \times a$ to zero. Choose a value of a that is a power of three for better results (this way, when we subdivide the square into boxes, the length of a side of the box remains an integer).

(ii) The elements of the array which are part of the box which is to be 'deleted' are set to one.

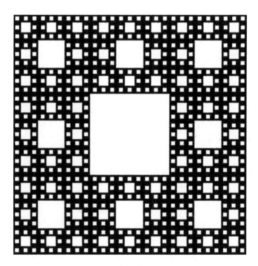

Figure 11.23. The Sierpinski carpet.

(iii) Once the box has been deleted, we display the array *A* using the 'imshow' command; 'imshow' displays the elements of an array in the MATLAB figure window. Once the array has been displayed, we pause the program using the 'pause' command. *Pause*(*x*) will pause the program for *x* seconds (*x* can also be a fraction of a second). The 'pause' command has to be used, otherwise the image of the carpet will change very fast and we will not be able to see the different stages of the carpet.

If you have implemented the above algorithm correctly, the successive images you see will be similar to those given in figure 11.24.

The boxes that have been deleted are 'white', while the boxes which are retained are shown in black. The result after the first iteration is shown at the top left corner. The result of second iteration is shown at the top right (in figure 11.24), while the results of the next two iterations are at the bottom left and bottom right, respectively.

```
'sierpinski carpet';
clear;
clf;

b=5;
blocksize=3^b;
for i=1:blocksize
    for j=1:blocksize
        a(i,j)=0;
    end
end
'initializing the array';


for k=0:b-2
    blocksize=round(blocksize/3);

    for i=1:3^k
        for j=1:3^k
            for n=blocksize:blocksize*2
                for m=blocksize:blocksize*2
                 a((i-1)*blocksize*3+n,(j-1)*blocksize*3+m)=1;
                end
            end
        end
    end
    imshow(a);
    pause(5);
end
```

The Von Koch curve, which resembles a snowflake is another well-known fractal. It is generated by starting with an equilateral triangle of side a. In the next iteration, the middle one-third of each side of the triangle is replaced by an equilateral triangle

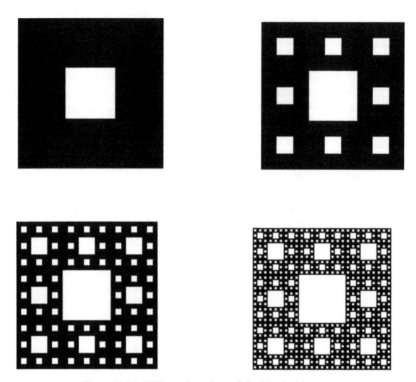

Figure 11.24. Different iterations of the Sierpinski carpet.

wide sides of $a/3$, as shown in the figure. This process is repeated many times over until a figure resembling a snowflake is generated.

While the fractals discussed above were artificially generated by recursively using a simple rule, fractals can also be seen in nature. For instance, if we examine a photograph of an entire island (figure 11.25) and calculate the length of the coastline, we get a certain value.

However, if we zoom into and magnify any part of the coastline (figure 11.26), we start to see finer features (bays and creeks) which were not visible earlier. Hence, the length of the coastline increases.

Continuing this process of zooming in will increase the length of the coastline further. Does this mean that the length of the coastline (or any fractal, such as the Koch curve) is infinite? In principle, the answer is affirmative. In practice, the recursive process has to stop once the atomic scale is reached.

The question about the length of the coastline is what started the field of fractals. *How Long Is the Coast of Britain? Statistical Self-Similarity and Fractional Dimension*[6] was one of the first papers published about fractals, written by Benoit Mandelbrot in 1967.

---

[6] *Science* **156**, 1967, 636–638. The paper can be accessed online at https://users.math.yale.edu/~bbm3/web_pdfs/howLongIsTheCoastOfBritain.pdf.

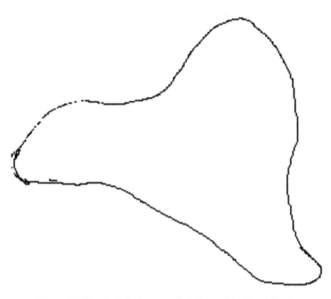

Figure 11.25. Aerial photograph (schematic) of an island.

Figure 11.26. Zooming in on a part of the coast. Smaller features are now visible.

We use the self-similar nature of fractals to define the dimensionality of a fractal. If a self-similar set (fractal) is composed of m copies of itself scaled down by a factor of r, the similarity dimension is defined by the equation:

$$m = r^d \tag{11.31}$$

or equivalently, $d = \frac{\ln m}{\ln r}$.

For the Cantor set, we have two copies remaining (straight line segments) when the straight line is divided into three pieces. Therefore, the dimensionality of the Cantor set is $d = \ln 2/\ln 3 = 0.63$. In general, the dimensionality of a fractal is not an integer (unlike regular geometrical figures, such as squares or cubes).

The study of fractals and chaos has really taken off in recent times, and it has been found that they also have practical applications. For example, the self-similar nature of fractals has been used in image compression. Since a small part of the fractal looks similar to the whole, it is sufficient to store or transmit a small part of the fractal.

We close this section on fractals by discussing the Mandelbrot set, which popularized the study of fractals. The Mandelbrot set is defined by the following map in the complex plane:

$$z = z^2 + c. \tag{11.32}$$

We start with $z = 0$ and some point c in the complex plane and iterate the above map; if z remains finite after an infinite number of iterations, point c is a member of the Mandelbrot set.

Figure 11.27 shows the Mandelbrot set. Points within the set are darkened. The centre of the figure corresponds to the origin of the complex plane. Points which correspond to $|c| < 2$ are mostly within the set, while points which correspond to $|c| > 2$ generally grow without bounds to infinity. Note the complex structure generated by a simple map (11.32). Note that the set has left-right symmetry, but is not symmetric about the centre. In addition, you can see the 'heart' within the set, which shows Mandelbrot's love for fractals! Note that the Mandelbrot set 'hangs' by a slender thread, which runs along the imaginary axis!

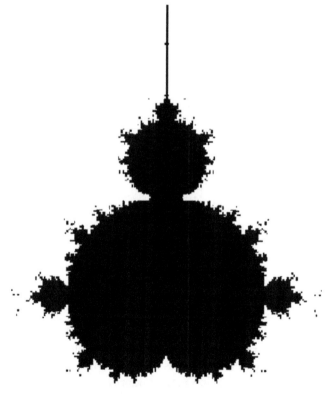

Figure 11.27. The Mandelbrot set.

Exercises

11.1. Consider the following one-dimensional tent map:
$f(x) = 2x$ for $0 \leqslant x \leqslant \frac{1}{2}$
$f(x) = 2(1-x)$ for $\frac{1}{2} \leqslant x \leqslant 1$.
What is/are the fixed points of this map? Determine the stability of the fixed points.

11.2. Consider the one-dimensional map $x_{n+1} = r^*\tanh(x_n)$, where r is a real number.
   (a) Initially, r is much smaller than one. What is/are the fixed point(s) of this map?
   (b) Determine the stability of the fixed point(s) mentioned above.
   (c) Draw the bifurcation diagram for this map, clearly indicating the value of r at which the first bifurcation occurs.
   (d) Determine the nature of the stability of each fixed point greater than the bifurcation point.

11.3. Explain why the bifurcation diagram for the sine map ($x_{n+1} = r\sin(\pi x_n)$) is similar to the bifurcation diagram of the logistic map.

11.4.
   (a) Find the approximate value of at least one fixed point (other than the one at $x = 0$) of $\dot{x} = e^x - \cos x$.
   (b) Determine the nature of the fixed point at $x = 0$.
   (c) How many fixed points are there in total for this equation?
   (d) Does this equation have any fixed point(s) for positive values of x? Why?

11.5. Consider the system $\dot{x} = rx - \sin x$.
   (a) How many fixed points does this system have for $r > 1$?
   (b) As r decreases from ∞ to zero, what is the first value of r at which bifurcation happens? What is the nature of this bifurcation?
   (c) If r is reduced further after the first bifurcation, subsequent bifurcations occur at smaller values of r. Obtain the equations satisfied by these bifurcation points as r is reduced towards zero.

11.6. What is the fixed point of the differential equation $\frac{dx}{dt} = \frac{x}{1+x^2}$? Suppose that you were using the Newton–Raphson method to determine the fixed point; what range of starting values of x would converge to the fixed point? Explain your answer.

11.7. Can the logistic map $x_{n+1} = rx_n(1 - x_n)$ be used to generate a random sequence of numbers? If yes, what would be the range of this sequence (the minimum and maximum values of x that are possible); also indicate the approximate value of r which should be used if the sequence has to be random.

11.8. Determine all the fixed points of the differential equation: $\frac{dx}{dt} = x^3 - 3x^2 + 2x$. Analyse the stability of these fixed points.

11.9. Determine all the fixed points of the map $x_{n+1} = \frac{2x_n}{1 + x_n}$. Analyse the stability of these fixed points.

Appendix A

Solutions to selected exercises

Chapter 2

2.4 We provide a flowchart, which should help you to write the program:

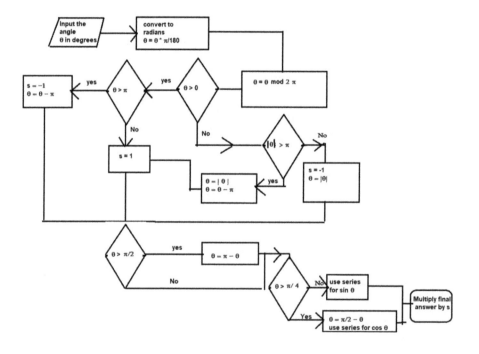

Chapter 3

3.1 Since $|\sin x| \leqslant 1$ for all x, a solution to the given equation exists if the right-hand side is less than or equal to one, i.e. $|x| \leqslant 1$. Therefore, $|x| \leqslant 10\pi$

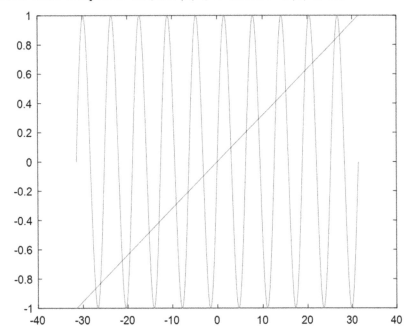

It is clear from the above graph that there are two solutions (points where $f(x) = x/10\pi$ intersects $g(x) = \sin x$) for a range of 2π. Therefore, for $|x| \leqslant 10\pi$, there will be nine solutions for positive values of x, nine solutions for negative values of x, and one zero solution. One of the solutions is between $x = \pi/2$ and $x=\pi$. Rewrite the equation as $\sin x - x/10\pi = 0$.

The next approximation to the solution is now the point at which the straight line connecting $(\pi/2, 19/20)$ and $(\pi, -1/10)$ meets the x-axis (via the method of false position). The equation for this line is:

$$\frac{y + \dfrac{1}{10}}{x - \pi} = \frac{\dfrac{19}{20} + \dfrac{1}{10}}{\dfrac{\pi}{2} - \pi} = \frac{-21}{10\pi}.$$

For $y = 0$, $x = \pi - \pi/21$. Substituting this into the expression $\sin x - x/10\pi$ gives us a positive value. Therefore, the root is between this value and π; thus, the equation for the straight line is modified to read:

$$\frac{y + \dfrac{1}{10}}{x - \pi} = \frac{0.0538 + \dfrac{1}{10}}{\pi - \pi/21 - \pi}.$$

Putting $y = 0$ gives the corresponding value of x, which is 3.0443. Verification: sin (3.0443)–(3.0443)/(10π) = 0.000 24, which is less than the value (0.0538) obtained in the previous iteration. This indicates the convergence of this method.

3.2 This equation will have only one real solution, since $x = \cos(x)$ at only one point, which happens when x is positive.

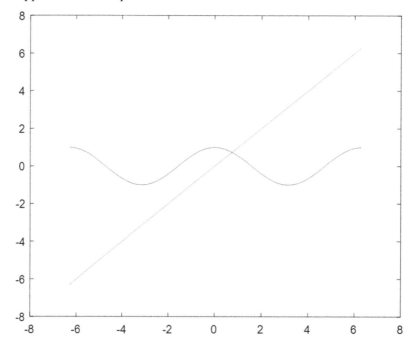

We can take the initial guess for the root to be $\pi/2$. By substituting this guess for x_i in the equation: $x_{i+1} = x_i - \dfrac{(\cos(x) - x)}{-\sin(x) - 1}$, we obtain $x = 0.7853$ and $x = 0.7395$ after the first and the second iterations, respectively.

3.3 The maximum value of x for which the function can be zero is when cos $(40x) = -1$. The corresponding value of x is $(1/40)^{1/4} = 0.3976$. Since the given function is an even function, we will have equal number of positive and negative zeros. Therefore, there are 12 zeros in total (two zeros for every 2π). The zeros occur within the range -0.3976 to $+0.3976$.

One of the zeros is between $x = 0$ (for which $f(x)$ is positive) and $x = \pi/40$ (for which $f(x)$ is negative).

By bisection, $c = (a+b)/2 = \pi/80$.

For $x = \pi/80$, $f(x)$ is positive, so the root will be between $x = \pi/40$ and $x = \pi/80$.

For the next iteration, $c = 3\pi/160$;

$f(3\pi/160)$ is negative, so the root is between $\pi/80$ and $3\pi/160$.

A graph of the function is shown below:

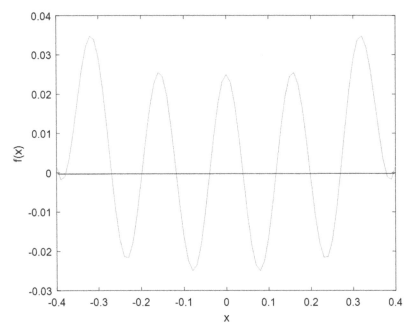

3.4 (a) According to Newton's second law, we have:

$$\frac{mv^2}{r} = k(r - r_o) = r - 1.$$

Therefore, $v = \sqrt{2}$ m s^{-1}.

(b) The angular momentum of the mass about the pivot $= mvr = l = 2\sqrt{2}$ Kg . m^2/s.

(c) The angular momentum will not be changed by the blow, as the tangential velocity is unchanged. The new energy of the mass is equal to $\frac{1}{2}m\dot{r}^2 + \frac{3}{2} = 2$, since energy is conserved in central force motion.

For $r_{\min}$ and $r_{\max}$, $E = U_{\text{eff}} = \frac{l^2}{2mr^2} + \frac{1}{2}(r - 1)^2 = 2.$

By substituting these values, we get $\frac{4}{r^2} + \frac{(r - 1)^2}{2} = 2.$

By rearranging the terms, we get $r^4 - 2r^3 - 3r^2 + 8 = 0.$

For the Newton–Raphson method, we have:

$$x_{i+1} = x_i - \frac{f(x_i)}{f'(x_i)} = r - \frac{(r^4 - 2r^3 - 3r^2 + 8)}{(4r^3 - 6r^2 - 2r)}.$$

We make an initial guess of $r = 1$, and substituting in the above expression, we get 1.5 and 1.45 in the first two iterations.

If we take $r = 3$ as the initial guess, we then get 2.77 and 2.71 in the first two iterations. Therefore, $r_{\min} = 1.45$ and $r_{\max} = 2.71$.

3.5 To determine the number of real solutions to the equation $4x^3 - 1 - \exp(-x^2) = 0$, we plot $4x^3 - 1$ and $\exp(-x^2)$ as two separate curves. The number of points at which these two curves intersect gives us the number of real solutions. From the plot given below, we can see that the two curves intersect at only one point (a positive value of x).

A-4

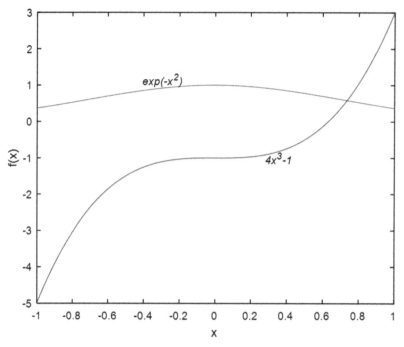

For $x = 0$, $4x^3 - 1 - \exp(-x^2) = -2$ and for $x = 1$, $4x^3 - 1 - \exp(-x^2) = 2.6321$. Therefore, $x = 0$ and $x = 1$ can be used as the two starting values of x for the bisection method.

After one iteration, we have to take one of the values of x to be 0.5. For $x = 0.5$, $4x^3 - 1 - \exp(-x^2) = -1.2788$.

Therefore, the solution lies between 0.5 *and* one;
so, after two iterations, the solution is $x = 0.75$.

3.6 A schematic of the situation described in this problem is as follows:

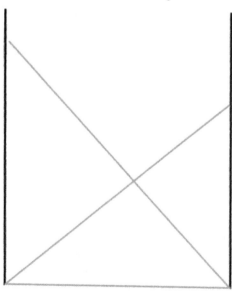

Let the width of the corridor be 'w'.

Choose the origin to be at the bottom of the left wall. Therefore, the equation for the straight line that represents the 20-ft ladder is:

$$y = \frac{\sqrt{20^2 - w^2}}{w} x. \tag{1}$$

The equation representing the other ladder is:

$$y = \frac{-\sqrt{30^2 - w^2}}{w} x + \sqrt{30^2 - w^2}. \tag{2}$$

The x coordinate of the point where the ladders cross each other is:

$$x = \frac{10w}{\sqrt{20^2 - w^2}}. \tag{3}$$

Using (1) and (2) and substituting (3) in (1) and (2), we obtain:

$$\frac{\sqrt{20^2 - w^2}}{w} \times \frac{10w}{\sqrt{20^2 - w^2}} = \frac{-\sqrt{30^2 - w^2}}{w} \times \frac{10w}{\sqrt{20^2 - w^2}} + \sqrt{30^2 - w^2}.$$

Simplifying this equation gives us:

$$\frac{-10\sqrt{30^2 - w^2}}{\sqrt{20^2 - w^2}} + \sqrt{30^2 - w^2} = 10$$

or

$$\frac{10\sqrt{30^2 - w^2}}{\sqrt{20^2 - w^2}} - \sqrt{30^2 - w^2} + 10 = 0. \tag{4}$$

Solving this equation will give us the width of the corridor.

(b) We see from equation (4) that w has to be less than 20, as otherwise we end up with imaginary numbers in equation (4).

Let us choose 10 and 15 to be the two initial guesses for w in equation (4).

Substituting $w=10$ into equation (4) gives us -1.9543 for the left-hand side.

Substituting $w = 15$ into equation (4) gives us $+3.6588$.

Therefore, these two values are valid guesses.

(c) Therefore, an approximate value for w is $(15+10)/2 = 12.5$ ft.

3.7 If the tank is filled with water to a height h, then the volume of water is given by:

$$\int_0^h \pi r^2 dh = \int_0^h \pi(R^2 - (R-h)^2)dh = \pi\left(Rh^2 - \frac{h^3}{3}\right) = \frac{\pi h^2}{3}(3R - h) = 30.$$

This equation can be rewritten as $f(h) = h^3 - 9h^2 + \frac{90}{\pi} = 0$ (since $R = 3$ m).

We see that $f(1) > 0$ and $f(3) < 0$.

Therefore, these two values of h can be used as starting values for root finding by the bisection method. According to the bisection method, the approximate value of the root in the next iteration is obtained using the equation: $c = (a+b)/2 = 1.5$.

Since $f(1.5) > 0$, the root is between 1.5 and 3. After proceeding in this manner for a few more iterations, the successive approximations to the root are: 2.25, 1.875, and 2.0625.

3.8 In the Newton–Raphson method, we look for solutions to the equation, $f(x) = 0$.
The condition for convergence is that $f'(x) \neq 0$.
In the case of the iterative method, we look for solutions to the equation $x = g(x)$, where $f(x) = g(x) - x$.
Here, the condition for convergence is that $|g'(x)| < 1$.
Now $f'(x) = g'(x) - 1$.
Therefore, if $g'(x) = 1$, neither method converges.

3.9 According to the Newton–Raphson method, the approximate values of the root at the ith and $(i+1)$th iteration are related by the following relation:

$$x_{i+1} = x_i - \frac{f(x_i)}{f'(x_i)} = x_i - \frac{(x_i^3 - N)}{(3x_i^2)} = \frac{2x_i}{3} + \frac{N}{3x_i^2}.$$

3.10 It is clear that the first equation is the equation of a circle of radius 3, and the second equation will intersect the circle at two points. We therefore expect two real values of x to satisfy the equation. We can use $x = 0$ as a starting value for the fixed-point iteration process.

To check whether the method converges, we rewrite the first equation as:

$$x = f(x) = \frac{1}{2} \ln\left(\sqrt{(9 - x^2)} \right).$$

Therefore, we have:

$$f'(x) = \pm \frac{-x}{\sqrt{9 - x^2}}.$$

The derivative is less than one for $x = 0$ or $x = 1$, therefore we expect the process to converge for these initial values. These starting values will converge to the correct root, which is in between these two values (0.5410). The negative root cannot be obtained by this method, as the result of the iteration is always a positive value.

3.11 Static equilibrium of the rods implies that they are in rotational and translational equilibrium.
Let the length of the spring at equilibrium be x m.
From the figure, we have the equation $2\cos\theta + x = 3$.
Consider the rotational equilibrium of one of the rods:
The torque due to the weight of rod (about the hinge) should be equal to the torque due to the spring force:

$(g\cos\theta)/2 = k(x-1)\sin\theta.$

Substituting for x in the above equation (from the first equation):

$(g\cos\theta) = 2k(3-2\cos\theta-1)\sin\theta$.

By substituting the values of g and k, we obtain the equation:

$$\tan\theta(1 - \cos\theta) - \frac{1}{20} = 0.$$

Choose $0°$ and $30°$ as the starting values of θ.

The left-hand side is negative for $0°$ and positive (equal to 0.027) for $30°$. Bisect the interval. For $\theta = 15°$, the left-hand side is negative. Therefore, after two iterations, the root is $22.5°$ (the exact value is $26.107°$).

3.12. To analyse the possible orbits under a central force, we should use the concept of the effective potential, which is given by:

$$U_{\text{eff}}(r) = \frac{l^2}{2mr^2} - U(r) = \frac{l^2}{2mr^2} - \frac{k}{r} - \frac{C}{r^3} - \frac{D}{r^4}.$$

Circular orbits correspond to minima in the effective potential.

Therefore, for circular orbits, $\frac{dU_{\text{eff}}}{dr} = \frac{-l^2}{mr^3} + \frac{k}{r^2} + \frac{3C}{r^4} + \frac{4D}{r^5} = 0$.

By taking the lowest common multiple and simplifying, we arrive at the following cubic equation:

$$\frac{dU_{\text{eff}}(r)}{dr} = 0 = \frac{-l^2r^2 + mkr^3 + 3mcr + 4mD}{mr^5}.$$

We get

$$mkr^3 - l^2r^2 + 3mcr + 4mD = 0.$$

We observe that this cubic equation has only two changes in the signs of the coefficients, so only two positive roots are possible. There can be one negative root, but since negative values of r are not possible, only two circular orbits of different radii are possible.

3.13 (a) The block will leave the platform when the normal reaction becomes zero.

Applying Newton's second law on the platform, we have: $-mg = -m\omega^2x$,

where x is the displacement of the block from the equilibrium position and w is the angular frequency of oscillation.

Substituting these values, we get $x = 2.5$ cm.

(b) The velocity of the block at this point will be equal to the velocity of the platform:

$$=5\omega\sin(\omega t) = 20\sqrt{5^2 - 2.5^2} = 86.6025 \text{ cm s}^{-1}.$$

(c) $x_b = 2.5 + ut - \frac{gt^2}{2} = 2.5+86.6025t - 500t^2$

$x_p = -5\cos\left(\omega t + \frac{2\pi}{3}\right)$. The phase factor of $2\pi/3$ is required as $t = 0$ when the block leaves the platform.

We need to determine the time at which $h(t) = x_p - x_b = 0$. We see $h(0.2)<0$ and $h(0.3)>0$. Therefore, the solution is between these two values. We apply bisection and find that $h(0.25) > 0$. So the root is between $t = 0.2$ and 0.25. $h(0.225)$ is negative, so the approximate root after two iterations is 0.2375.

3.14 (a) For $x < 0$, the solution will be of the form:

$$\psi_I(x) = A \exp(\alpha x),$$

where $\alpha = \sqrt{\dfrac{2m(V_o - E)}{\hbar^2}}$.

For $0 < x < a$, the solution is $\psi_{II}(x) = B \sin(\beta x) + C \cos(\beta x)$,

where $\beta = \sqrt{\dfrac{2mE}{\hbar^2}}$.

For $x > 0$, $\psi(x) = 0$

(b) By applying the boundary conditions to the wavefunction and the derivative, we obtain:

$$A = C.$$

After we apply the condition that the derivative has to be continuous, we have:

$$\alpha A = B\beta.$$

If we also apply the condition that the wavefunction vanishes at $x = a$, we obtain:

$$\psi_{II}(a) = B \sin(\beta a) + C \cos(\beta a) = 0.$$

3.15 (a) $x^7 + 3x^5 + 4x^3 + 5x = x(x^6 + 3x^4 + 4x^2 + 5)$

$x = 0$ is a root of this equation. There is no change of sign in the polynomial within brackets, so there are no positive real roots. Replacing x with $-x$ also leads to a polynomial with no change of sign and therefore no negative real roots. Zero is the only real root.

(b) $x = n\pi$ are the real solutions of this function, as $\sin(\sin(n\pi)) = \sin(0) = 0$. Therefore, there are infinite solutions.

(c) $\sin(\cos(x)) = 1$

$\cos(x) \leqslant 1$ and $\sin(1 \ rad) < 1$; therefore; there is no real solution.

(d) $\tanh(x^2) = \dfrac{\exp(x^2) - \exp(-x^2)}{\exp(x^2) + \exp(-x^2)}$ $\tanh(x^2) = 0$ if $\exp(x^2) = \exp(-x^2)$, which happens only at $x = 0$.

3.16 All the terms of the given polynomial have even powers of x and the constant term is also positive.

Therefore, the left-hand side of the above equation is always positive and never zero. Therefore, this equation does not have any real solutions. Since this is an eighth-degree polynomial, it will have eight solutions in total (four complex conjugate pairs).

3.17. For the bisection method to work, we need two starting values, a and b, such that $f(a)*f(b) < 0$,

i.e. the function should intersect the x-axis at the root. However, if the function just touches the x-axis at the root, i.e., $f(x_o) = 0$ and $f'(x_o) = 0$, both the function and its derivative are zero at x_o; if this is the case, the bisection method cannot be used to determine the root. This happens, for example, for the equation $x^2 - 2x + 1 = 0$.

3.18 Bisection will not work for complex roots, as we are bisecting the interval in the bisection method. Complex roots lie on the complex plane. Therefore, we should use the Newton–Raphson method to determine complex roots.

3.19 According to the Newton–Raphson method, the approximate values of the root at the ith and $(i+1)$th iteration are related by the following relation:

$$x_{i+1} = x_i - \frac{f(x_i)}{f'(x_i)} = x_i - \frac{(x_i^3 - N)}{(3x_i^2)} = \frac{2x_i}{3} + \frac{N}{3x_i^2}.$$

Chapter 4

4.1 The polynomial of degree n that passes through a given set of $n+1$ points is unique. Therefore, both $L_n(x)$ and $P_n(x)$ correspond to the same polynomial. $L_n(x)$ and $P_n(x)$ are two different forms of the same polynomial. Therefore, the error will be same, regardless of whether we use $L_n(x)$ or $P_n(x)$.

4.2 (i) The series expansion $\exp(x) = 1 + x + x^2/2$ is valid only if $|x| \ll 1$. Therefore, the series expansion is not valid if $|x|$ is greater than one.

(ii) Furthermore, the Lagrange polynomial matches $\exp(x)$ at three points, whereas the series expansion only matches $\exp(x)$ at $x = 0$.

Therefore, the value determined by the Lagrange interpolation polynomial will be more accurate.

4.3.

$$\ln(9.3) = \left(\frac{9.3 - 9.5}{9 - 9.5}\right)2.1972 + \left(\frac{9.3 - 9}{9.5 - 9}\right)2.2513 = 2.22966$$

The error is given by $\frac{(x - x_0)(x - x_1)}{3!}\frac{d^2 f}{dx^2}$.

Here, $f(x) = \ln(x)$

Therefore, the error in this case is approximately: $-\frac{(9.3 - 9)(9.3 - 9.5)}{3!}\frac{1}{x^2}$.

By substituting 9 and 9.5 for x, we find that the error is in the range from 0.000 33 to 0.000 37.

We also find that the exact error is $\ln(9.3) - 2.229\ 66 = 0.000\ 35$.

4.4

$$L(x) = \frac{(x - 4)(x - 3)(x - 8)}{(2-4)(2-3)(2-8)} + 3\frac{(x - 2)(x - 3)(x - 8)}{(4-2)(4-3)(4-8)}$$
$$+ 5\frac{(x - 4)(x - 2)(x - 8)}{(3-4)(3-2)(3-8)} + 9\frac{(x - 4)(x - 3)(x - 2)}{(8-4)(8-3)(8-2)}$$

$$L(x) = -\frac{(x - 4)(x - 3)(x - 8)}{12} - 3\frac{(x - 2)(x - 3)(x - 8)}{8}$$
$$+ \frac{(x - 4)(x - 1)(x - 8)}{1} + 9\frac{(x - 4)(x - 3)(x - 2)}{120}$$

At $x = 5$, $L(x) = -1.3$.

4.5 The forward difference formula assumes that the function is approximated by a straight line between x_1 and the previous point. This is obviously inferior to approximating it by a polynomial of degree greater than two. To illustrate this, let the table for $f(x)$ versus x be as given below:

x	$f(x)$
0	0
0.1	0.001
0.2	0.008
0.3	0.027
0.4	0.064
0.5	0.125

If $x_1 = 0.2$ and we require the derivative at this point, the forward difference formula will give us the value 0.007. We can fit the given data exactly to a cubic, $f(x) = x^3$. Therefore, the derivative of this will be $3x^2$. At x_1, the value of the derivative will be 0.12. This value is exactly correct, while the forward difference formula is only an approximation (quite a poor approximation in this case).

Chapter 5

5.1 If the initial vector is perpendicular to A, it will not converge to A, as there is no component of that vector in the direction of A (see the proof of power method). A possible choice is $[-1 \ -1 \ 1]$.

5.2 By applying the transformation $B = A - \lambda_1 v_1 x^T$, we reduce the largest eigenvalue λ_1 to zero. The determinant of a matrix is equal to the product of its eigenvalues. Therefore, the eigenvalue of B will be equal to zero. Thus, if the eigenvalues of matrix A were all nonzero, the determinant of B will not be equal to the determinant of A.

5.3(a) Subtract the second row of the determinant from the first row and replace the first row by the result; this yields:

$$\begin{vmatrix} 0 & x & 0 & \cdots & 0 \\ 1 & 1-x & 1 & \cdots & 1 \\ 1 & 1 & 2-x & \cdots & 1 \\ \cdots & \cdots & \cdots & \cdots & \cdots \\ 1 & 1 & 1 & \cdots & n-x \end{vmatrix} = -x \begin{vmatrix} 1 & 1 & 1 & \cdots & 1 \\ 1 & 1 & 2-x & \cdots & 1 \\ 1 & 1 & & \cdots & 1 \\ \cdots & \cdots & \cdots & \cdots & \cdots \\ 1 & 1 & 1 & \cdots & n-x \end{vmatrix}.$$

Proceeding in a similar fashion, we see that the determinant is reduced to $(-1)^{n-1} x(x-1)(x-2)\ldots (x-n-1)$.

(b) To evaluate the above expression, we need to perform $n-1$ additions and $n-1$ multiplications. This is much smaller than the number used in the direct evaluation of the determinant, which will take $O(n^n)$ additions and multiplications.

(c) If the determinant is of size 4×4 and $x = 5$, the determinant is equal to $(-1)^3 * 5 (5-1)(5-2) = -60$.

(d) The diagonal elements need to be changed to generate the required matrix. Therefore, we need to add the following lines of code:

for i=2:n
A(i,i)=i x;
end.

5.4 (a)

$$
\begin{pmatrix} a_{11} & a_{12} & a_{13} & .... \\ a_{21} & a_{22} & a_{23} & .... \\ a_{31} & a_{32} & a_{33} & .... \\ a_{41} & a_{42} & a_{43} & .... \end{pmatrix} = \begin{pmatrix} 1 & 0 & 0 & .... \\ l_{21} & 1 & 0 & .... \\ l_{31} & l_{32} & 1 & .... \\ l_{41} & l_{42} & l_{43} & .... \end{pmatrix} \begin{pmatrix} u_{11} & u_{12} & u_{13} & .... \\ 0 & u_{22} & u_{23} & .... \\ 0 & 0 & u_{33} & .... \\ 0 & 0 & 0 & .... \end{pmatrix}
$$

From the above matrix equation, we see that $u_{1k} = a_{1k}$.

(b) $l_{k1} = \dfrac{a_{k1}}{u_{11}} = \dfrac{a_{k1}}{a_{11}}$.

(c) $u_{ik} = a_{ik} - \sum_{j=1}^{i-1} l_{ij} u_{jk} \quad k = i, i+1, i+2 ..., N$

$$
l_{kj} = \dfrac{a_{kj} - \sum_{i=1}^{j-1} l_{ki} u_{ij}}{u_{jj}} \quad k = 1, 2, 3, , , , j^{-1}.
$$

5.5 When we multiply two $N \times N$ matrices, we get a matrix with N^2 elements. To calculate each of these elements, we require N multiplications. Therefore, the total number of multiplications is N^3. The generation of each element of the product matrix requires $N-1$ additions. Since there are N^2 elements in total, we require $N^2(N-1)$ additions.

If $N = 100$, the total time taken will be $N^2(N-1)+N^3 = 99 \times 10^4 + 10^6 = 1\ 990\ 000$ nanoseconds = 1 990 microseconds.

5.6 (a) The equations of motion for the three masses will be:
for the topmost mass: $m\ddot{x}_1 = -kx_1 + k(x_2 - x_1)$,
for the middle mass: $m\ddot{x}_2 = k(x_3 - x_2) - k(x_2 - x_1)$,
for the lowermost mass: $m\ddot{x}_3 = -k(x_3 - x_2)$.
In these equations, the downward direction is positive.

(b) By substituting the solution into these equations, we obtain:

$$
-\frac{2k}{m}A_1 + \frac{k}{m}A_2 = -\omega^2 A_1
$$

$$
\frac{k}{m}A_1 - \frac{2k}{m}A_2 + \frac{k}{m}A_3 = -\omega^2 A_2
$$

$$
\frac{k}{m}A_2 - \frac{k}{m}A_3 = -\omega^2 A_3.
$$

These three equations can be written in the form of an eigenvalue/matrix equation:

$$\begin{pmatrix} 20 & -10 & 0 \\ -10 & 20 & -10 \\ 0 & -10 & 10 \end{pmatrix} \begin{pmatrix} A_1 \\ A_2 \\ A_3 \end{pmatrix} = \omega^2 \begin{pmatrix} A_1 \\ A_2 \\ A_3 \end{pmatrix}.$$

We start with (1 1 1) as the eigenvector. After two iterations, the eigenvalue is 20 (therefore, $\omega \sim 20^{0.5}$).

5.7 From (5.40), we see that to determine the elements of the upper triangular matrix U, we need to divide by the diagonal elements of the L matrix. If the diagonal elements of the L matrix are zero, we get a 'divide by zero' error.

5.8 No. The matrices L and U do not commute. We know that $l_{11}u_{11} = a_{11}$. But if we consider the product UL, the first element of the product will be:

$$\sum_{i=1}^{n} u_{1i}l_{i1} \neq l_{11}u_{11}.$$

Therefore, the matrices L and U do not commute.

5.10. (b) The slope and intercept of the best-fit line turn out to be −0.012 6124 and 35. 011 581, respectively. Accordingly the 'predicted' winning time for the 2021 Olympics is 9.53 s.

(c) Obviously, the linear downward trend cannot continue forever. We expect the decrease to saturate at some point in time and reach a constant value.

5.11. The graph of thickness versus breaking strength looks like this on a logarithmic scale:

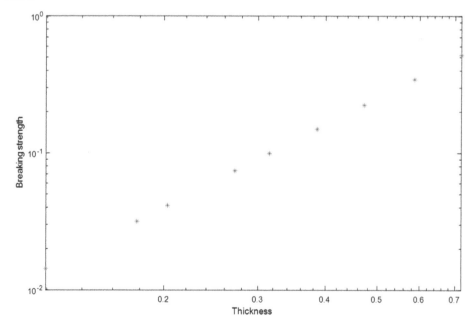

It fits a straight line with a slope of two, i.e. the breaking strength is proportional to the square of the thickness.

5.12 (a) The eigenvalues are simply the diagonal elements: 1, 2, 3, and 4.

(b) To find the eigenvectors, substitute the eigenvalues into the eigenvalue equation. For the smallest eigenvalue, we have:

$$\begin{pmatrix} 1 & 0 & 0 & 0 \\ 6 & 2 & 0 & 0 \\ 7 & 4 & 3 & 0 \\ 5 & 6 & 7 & 4 \end{pmatrix} \begin{pmatrix} x_1 \\ x_2 \\ x_3 \\ x_4 \end{pmatrix} = \begin{pmatrix} x_1 \\ x_2 \\ x_3 \\ x_4 \end{pmatrix}.$$

This leads to the following system of equations:
$6x_1 + x_2 = 0,$
$7x_1 + 4x_2 + 2x_3 = 0,$
$5x_1 + 6x_2 + 7x_3 + 3x_4 = 0.$
If $x_1 = 1$, we have $x_2 = -6$, $x_3 = 17/2$, $x_4 = -19/2$.

Therefore, the eigenvector corresponding to the smallest eigenvalue is: $\begin{pmatrix} 1 \\ -6 \\ 17/2 \\ -19/2 \end{pmatrix}$.

For the largest eigenvalue, we get the following equations:
$x_1 = x_2 = x_3 = 0.$

Therefore, the corresponding eigenvector is $\begin{pmatrix} 0 \\ 0 \\ 0 \\ 1 \end{pmatrix}$.

(c) The determinant = the product of the eigenvalues = 24.

Chapter 6

6.1 (a) Since $\cos(x)$ is always less than or equal to one, $\dfrac{\cos x}{1 + x^2} \leqslant \dfrac{1}{1 + x^2}$

$\displaystyle\int_1^\infty \frac{\cos x}{1 + x^2}\,dx \leq \int_1^\infty \frac{1}{1 + x^2}\,dx = \tan^{-1} x\big|_1^\infty = \frac{\pi}{4}.$ Therefore, the given integral converges.

(b) We can use the change of variables $x = 1/y$, which will transform the integral

into: $\displaystyle\int_1^\infty \frac{\cos x}{1 + x^2}\,dx = -\int_1^0 \frac{\cos\left(\dfrac{1}{y}\right)}{1 + \left(\dfrac{1}{y}\right)^2}\,\frac{dy}{y^2}.$

(c) Now change from y to t such that t goes from -1 to 1.
$y = (t+1)/2$:

$$\int_1^\infty \frac{\cos x}{1+x^2}dx = \int_{-1}^1 \frac{\cos\left(\dfrac{2}{t+1}\right)}{1+\dfrac{(t+1)^2}{4}}dy$$

$$= \int_{-1}^1 \frac{2\cos\left(\dfrac{2}{t+1}\right)}{4+(t+1)^2}dt = 2\left[\frac{\cos\left(\dfrac{2}{1+\dfrac{1}{\sqrt3}}\right)}{4+\left(1+\dfrac{1}{\sqrt3}\right)^2} + \frac{\cos\left(\dfrac{2}{1-\dfrac{1}{\sqrt3}}\right)}{4+\left(1-\dfrac{1}{\sqrt3}\right)^2}\right] = -0.1013.$$

6.2 The length of a small portion of a curve is given by:
$ds = \sqrt{dr^2 + r^2 d\theta^2}$. From the given equation for the curve, we have
$dr = 2\cos(2\theta)d\theta$.
The total length of the curve is given by the integral:

$$\int_0^{2\pi} \sqrt{4\cos^2 2\theta + \sin^2 2\theta}\,d\theta = \int_0^{2\pi} \sqrt{1+3\cos^2 2\theta}\,d\theta = \int_0^{2\pi} \sqrt{\frac{5}{2} + \frac{3\cos 4\theta}{2}}\,d\theta. \quad (10)$$

To use Gaussian quadrature, we make a variable transformation from θ to ϕ, so that the limits for ϕ are from -1 to 1.

The transformation is defined by: $\dfrac{\theta}{2\pi} = \dfrac{\varphi+1}{2}$ or $\theta = \pi(\varphi+1)$:

$$d\theta = \pi d\varphi.$$

Therefore, the integral to be evaluated is transformed to
$$\pi\int_{-1}^1 \sqrt{\frac{5}{2} + \frac{3}{2}\cos[\pi(\varphi+1)]}\,d\varphi.$$
Using a two-point Gaussian quadrature, we get:

$$\pi\left[\sqrt{\frac{5}{2} + \frac{3}{2}\cos\left[\pi\left(1 - \frac{1}{\sqrt3}\right)\right]} + \sqrt{\frac{5}{2} + \frac{3}{2}\cos\left[\pi\left(1 + \frac{1}{\sqrt3}\right)\right]}\right] = 10.6276.$$

6.3 (a) The program is trying to compute the integral.

$$\int_0^{1000} \frac{x^{3/2}}{1+\exp(x-30.7333)}dx \text{ using Simpson's 1/3 rule.}$$

(c) The value of x is not being incremented inside the while loop, so the program will never terminate.

(d) The following commands have to be included after line number 12 (within the while loop) for the program to give the correct output:

$x0 = x0+2*h;$

$x1=x0+h;$

$x2=x0+2*h;$.

6.4 The domain of integration is finite, so the integral will converge if the integrand is finite at all points within the domain of integration.

At $x = 0$, the denominator goes to zero, so we have to check the limiting value of the function as $x \to 0$:

$$Ltx \to 0 \frac{\exp(-x) - 1 + x}{x} = \frac{\left(1 - x + \frac{x^2}{2} - 1 + x\right)}{x} = \frac{x}{2} \to 0.$$

Therefore, the given integral will converge.

To evaluate the integral by Gaussian quadrature, make of change of variables from x to t such that

$$\frac{x}{1} = \frac{t+1}{2} \text{ and } dx = dt/2.$$

Therefore, the given integral equals $\dfrac{f\left[-\dfrac{1}{\sqrt{3}}\right] + f\left[\dfrac{1}{\sqrt{3}}\right]}{2}$,

where $f[t] = \dfrac{\exp\left[-\dfrac{t+1}{2}\right] - 1 + \left(\dfrac{t+1}{2}\right)}{\left(\dfrac{t+1}{2}\right)}$.

After substitution, the value of the integral is found to be 0.2034.

6.5 In the range $0 \leqslant x \leqslant \pi/4$

$\sin(x) \leqslant \cos(x)$ (equality holds only for $x = \pi/4$).

Therefore, the integrand of $I_1 \geqslant I_2$ over the entire range.

Therefore $I_1 > I_2$.

6.6 $n = $ input('enter the number of points');

for $i=1:n$

for $j=1:n$

$A(i,j)= (j-1)^{(i-1)}$

end

end

for $i=1:n$

$A(i,n+1) = (n-1)^i/i$

end.

6.7 Using a comparison test, we have:

$\displaystyle\int_0^\pi \frac{d\theta}{\sqrt{\theta} + \sin\theta} < \int_0^\pi \frac{d\theta}{\sqrt{\theta}}$, as $\sin\theta$ is positive throughout the domain of integration.

$\displaystyle\int_0^\pi \frac{d\theta}{\sqrt{\theta}}$ converges as $Lta \to 0 \displaystyle\int_a^\pi \frac{d\theta}{\sqrt{\theta}} = 2(\sqrt{\pi} - \sqrt{a}) = 2\sqrt{\pi}$.

Therefore, the given integral converges, and we have
$$\int_0^\pi \frac{d\theta}{\sqrt{\theta} + \sin\theta} < \int_0^\pi \frac{d\theta}{\sqrt{\theta}} = 2\sqrt{\pi}.$$

6.8 To evaluate the integral by Gaussian quadrature using Legendre polynomials, the limit of integration needs to be −1 to +1.

Let y range from -1 to +1, while x ranges from $-\pi$ to $+\pi$.

The relation between these variables is given by:

$$x = \frac{2\pi y}{2} = \pi y$$

$$dx = \pi dy.$$

Therefore, the given integral can be rewritten as:

$$I = \int_{-1}^1 \frac{\sin(\pi y)}{\pi y}\pi dy = \int_{-1}^1 \frac{\sin(\pi y)}{y}dy.$$

As per the two-point Gaussian quadrature formula:

$$I = f(-1/\sqrt{3}) + f(1/\sqrt{3}) = \frac{2\sin(\pi*0.5773)}{0.5773} = 3.3623.$$

6.9 The Legendre seventh-degree polynomial will have seven roots, but we know that a Legendre polynomial of odd degree will always have zero as one of roots. Furthermore, since the Legendre polynomials are polynomials in $\cos(\theta)$, they are always even functions and therefore the roots will be symmetric with respect to the origin. Therefore, leaving out zero, we have six other roots; using the symmetry property, we see that it is enough the run the program thrice to get all the roots.

6.10 The exact value of the integral is zero, as the integrand is an odd function. Since the two-point Gaussian quadrature will give the exact result for polynomials up to degree three, we will get zero as the result.

6.11 (a) Substitute $x = t^2$;
$$\int_0^2 \frac{dx}{(1+x)\sqrt{x}} = \int_0^{\sqrt{2}} \frac{2dt}{(1+t^2)} = 2\tan^{-1}t \,|_0^{\sqrt{2}}.$$ Therefore, the integral converges.

(b) To evaluate this using Gaussian quadrature, we make the following change of variables:
$$\frac{x}{2} = \frac{t+1}{2}.$$ Therefore, the given integral becomes

$$\int_0^2 \frac{dx}{(1+x)\sqrt{x}} = \int_{-1}^1 \frac{dt}{(2+t)\sqrt{t+1}}.$$

Therefore, the given integral equals $f\left[-\frac{1}{\sqrt{3}}\right] + f\left[\frac{1}{\sqrt{3}}\right]$,

where $f[t] = \frac{1}{(t+2)\sqrt{t+1}}$.

On substitution, the value of the integral is found to be 1.3901.

Chapter 7

7.1 (a) The value of x varies from zero to two, while the integrand varies from zero to four. Therefore, the area of the smallest bounding rectangle is eight.

(b) In the hit-or-miss method, the integral is equal to $A*n_s/n$, where A is the area of the bounding rectangle (equal to eight in this case), while n_s is the number of hits and n is the total number of points used to evaluate the integral. If the integral is to be evaluated up to an accuracy of four decimal places, then we need to have a total of around 80 000 points.

The above figure is only a lower bound on the error. The actual error will be greater and can be estimated more exactly for the sample mean method.

For the sample mean method, the error (standard deviation) of the result is given by: $\sigma_m = \dfrac{\sigma}{\sqrt{n}}$, where σ is the standard deviation of the integrand from its mean value, and n is the number of points used to evaluate the integral.

The mean value of the function over the range of integration is:

$$\frac{1}{2}\int_0^2 x^2 dx = \frac{4}{3}$$

$$\sigma^2 = \langle f^2 \rangle - \langle f \rangle^2 = \frac{1}{2}\int_0^2 x^4 dx - \frac{16}{9} = \frac{32}{10} - \frac{16}{9} = \frac{128}{90} = \frac{64}{45}.$$

Therefore, to obtain an accuracy of 0.0001, the number of points that has to be used is approximately 2.0227×10^8.

7.2 (a) b should be a prime number for it to have a long period.

Otherwise, the sequence could lead to a zero value after some iterations (for example, if $x_n = 2$, $a = 5$, and $b = 32$).

b) Yes, interchanging would help. Only certain values from the set $\{0,1,2,...,b\}$ will produce all possible elements within this set when exponentiated. Therefore, it is better to keep x_n as the exponent. Since a is kept fixed, it can be chosen carefully to yield a long period.

For example, $b = 11$ and $a = 2$.

Different powers of 2 (mod 11) give rise to all possible values less than 11:
$2^1 = 2$, $2^2 = 2$, $2^3 = 8$, $2^4 = 11$, $2^5 = 10$, $2^6 = 9$, $2^7 = 7$, $2^8 = 3$, $2^9 = 6$, and $2^{10} = 1$.

If we have raised x_n to the power a, then since x_n is chosen randomly, it might be chosen to be a number that produces an output of one. The subsequent numbers will also be one. In the other case, even if the output is one, in the next iteration, it will be replaced by a, and in fact, the period is longest in this case.

7.3 First, $p(x)$ is the probability density function, whereas $p(x)dx$ is the probability that a given value of x lies in the range x to $x+dx$. We know that probability is equal to the frequency of occurrence of a certain value. Therefore, to determine the frequency, we need to decide a value for dx and then multiply it by $p(x)$; $p(x)dx = n/N$, where n is the number of values of x which lie between x and $x +dx$, and N is the total number of values of x we are generating. Therefore, we take a value of x, substitute it into $A\exp(-x^2)$ and multiply it by N to get n. By definition, n has to be an integer, whereas our procedure does not guarantee that it will be an integer, which means that we have to round it off, leading to some error in generating the distribution. Furthermore, these n values have to be uniformly distributed within the interval dx, which means we are not exactly generating the Gaussian distribution.

7.4 We can choose $p(x) = A\exp(-x)$, since it is similar to the integrand.
   A is found to be equal to $1/(1-\exp(-p))$ by normalization.

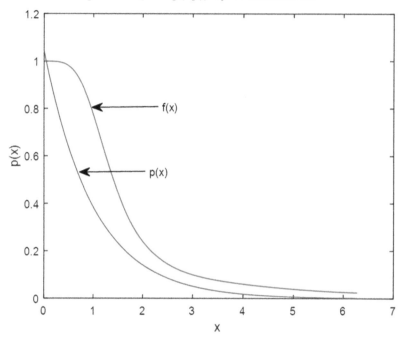

7.5 The sequence of numbers generated by this procedure would not be considered to be random. The number N_1 is of the form $10x+y$ and the number N_2 is of the form $10y+x$. Therefore, $N_3 = 9(x-y)$, and the subsequent numbers generated will only be multiples of nine. Furthermore, the sequence generated will have a very short period. For example, if the starting value is 17, the sequence will be 17, 54, 09, 81, 63, 27, 45, 09,....

7.6 Program
```
pt=1;
While pt < 101
r = rand;
x(pt)=2*(r−0.5);
if x(pt)>0
r1=rand;
y(pt)=r1*(1−x(pt));
else
y(pt)=r1*(1+x(pt))
end
pt=pt+1;
end.
```

7.7 The curves intersect at (0,0) and (9,3).

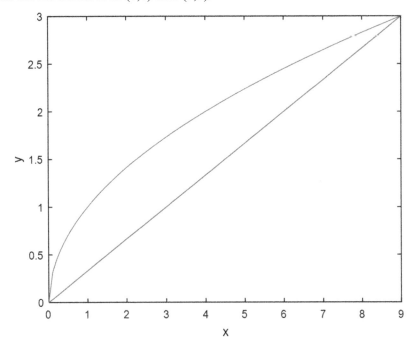

The area of overlap is the area between the curves.
The following MATLAB code can be used to evaluate the area of overlap:

```
hit=0;
N=1000 000;
For i=1:N
r=rand;
xr = 3*rand;
r=rand;
yr=9*r;
y1=xr^0.5;
y2=x/3;
if yr<y1 && yr>y2
hit=hit+1
end
end
integral=27*hit/N.
```

7.8 The base of the bounding rectangle will be equal to π.
To determine the other side of the minimum bounding rectangle, we need to find the maximum value of the integrand $f(x)$:

$$\frac{df}{dx} = -\exp(-x)\sin(x) + \exp(-x)\cos x = 0.$$

The maximum occurs when $\tan x = 1$ or $x = \pi/4$.

The value of the integrand at this point is 0.3224.

Therefore, the area of the minimum bounding rectangle is 1.0128.

7.9 (a) A is obtained by normalizing the probability distribution.

$$\int_{-1}^{1} p(x)dx = \int_{-1}^{1} \frac{A}{1 + x^2} dx = A\tan^{-1}x \big|_{-1}^{1} = \frac{A\pi}{2} = 1 \text{ or } A = \frac{2}{\pi}.$$

(b) While making the change of variables from r to x, we have to satisfy the relation:

$$\int_{-1}^{x} p(x)dx = \int_{0}^{r} p(r)dr = r$$

or $\dfrac{2}{\pi}\displaystyle\int_{-1}^{x} \dfrac{dx}{1 + x^2} = r$ or $\dfrac{2}{\pi}[\tan^{-1}x - \tan^{-1}(-1)] = r.$

Therefore: $x = \tan\left(\dfrac{\pi r}{2} - \dfrac{\pi}{4}\right).$

7.10 (a) The value of A is obtained by normalization.

$\displaystyle\int_{0}^{1} A(1 - x)dx = 1$. This gives $A = 2$.

(b) According to the inverse transform method,

$$r = \int_{0}^{x} 2(1 - x)dx = 2x - x^2;$$

$$\therefore x = 1 - \sqrt{1 - r}.$$

The other root, i.e. $x = 1 + \sqrt{1 - r}$, is not considered, since x has to be in the range zero to one.

7.11 The equation $9x^2 + 25y^2 = 225$ represents an ellipse centered at the origin with $a = 5$ and $b = 3$.

The equation $x^2 + y^2 - 10x + 16 = 0$ can be rewritten as $(x-5)^2 + y^2 = 9$, which shows that it represents a circle with a radius of three centered at $(5,0)$.

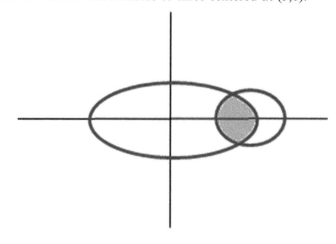

The region of overlap is shown shaded in the above figure.

Multiply the equation for the circle by 25 and subtract the other equation from it to obtain the following equation:

$16x^2-250x+625 = 0$

The points of intersection of the circle and the ellipse are at $x = 3.125$ and $x = 12.5$.

Only the first value of x is valid, since the other value gives an imaginary value for y when substituted into the equation for the circle/ellipse.

The corresponding value of y is $\sqrt{-x^2 + 10x - 16} = \pm2.342$.

Therefore, the limits for the x coordinate of the overlap region are $x = 2$ to $x = 5$.

The y coordinate of the bounding rectangle will be from $y = 0$ to $y = 2.35$.

However, the overlap region has to be broken up into two parts, as it is governed by the equation of the circle from $x = 2$ to $x = 3.125$ and the equation of the ellipse from $x = 3.125$ to $x = 5$.

MATLAB code:

```
hit=0;
N=1000 000;
For i=1:N
r=rand;
xr = (1.125*r)+2;
r=rand;
yr=r*2.35;
y=(-xr^2+10*xr-16)^0.5;
if yr<y
hit=hit+1
end
end
int1=(1.125*2.35)*hit/N;
hit=0;
For i=1:N
r=rand;
xr = (1.875*r)+3.125;
r=rand;
yr=r*2.35;
y=((225-*xr^2)/625)^0.5;
if yr<y
hit=hit+1
end
end
int2=(1.875*2.35)*hit/N
int=2*(int1+int2);
```

Chapter 9

9.1 (a) $x(t = 2) = x(t = 1) + \left.\dfrac{dx}{dt}\right|_{t=1} + \left.\dfrac{1}{2}\dfrac{d^2x}{dt^2}\right|_{t=1} + \left.\dfrac{1}{6}\dfrac{d^3x}{dt^3}\right|_{t=1} + \dots$

(b) $\dfrac{dx}{dt} = 1 + \dfrac{x}{t}$; differentiating this equation with respect to time:

$$\frac{d^2x}{dt^2} = \frac{1}{t}\frac{dx}{dt} - \frac{x}{t^2} = \frac{1}{t}\left(1 + \frac{x}{t}\right) - \frac{x}{t^2} = \frac{1}{t}$$

$\dfrac{d^3x}{dt^3} = -\dfrac{1}{t^2}\dfrac{d^4x}{dt^4} = \dfrac{2}{t^3}$.

Generalizing from the above, we have: $\dfrac{d^nx}{dt^n} = (-1)^n\dfrac{(n-2)!}{t^{n-1}}$, which is valid for $n \geqslant 2$.

(c) Therefore, the Taylor series now becomes:

$x(t = 2) = 1+2 + \dfrac{1}{1\times2} - \dfrac{1}{2\times3} + \dfrac{1}{3\times4} - \dots = 3 + \displaystyle\sum_{n=1}^{\infty}\frac{(-1)^{n+1}}{n(n+1)}$.

9.2. In the shooting method, we assume or guess the value of the first derivative at one boundary point and determine the corresponding value at the other boundary point. We determine the value at the second boundary corresponding to two different guesses for the first derivative and then use linear interpolation to find the solution that satisfies the given boundary conditions.

This method can easily be implemented for one-dimensional problems, but in higher dimensions, we can start from a given point and proceed in multiple directions until we reach a boundary point. There are an infinite number of boundary points and even if we were to discretize the boundary, the number of possible paths would still be too large, and therefore the shooting method would be very arduous for two-dimensional problems.

9.3 $x'(\pi) = -1$

a) By the Taylor series method:

$$x(2\pi) = x(\pi) + \left.\pi\frac{dx}{dt}\right|_{x=\pi} + \left.\frac{\pi^2}{2}\frac{d^2x}{dt^2}\right|_{x=\pi} + \left.\frac{\pi^3}{6}\frac{d^3x}{dt^3}\right|_{x=\pi} + \left.\frac{\pi^4}{4}\frac{d^4x}{dt^4}\right|_{x=\pi} + \dots$$

$$\frac{d^2x}{dt^2}(t = 2\pi) = \cos t - x' = -1+1=0$$

$$\frac{d^3x}{dt^3}(t = 2\pi) = -\sin t - x'' = 0$$

$$\frac{d^4x}{dt^4}(t = 2\pi) = -\cos t - x''' = 1$$

Substituting these values of the derivatives into the Taylor series gives us:

$$x(2\pi) = x(\pi) - \pi + \frac{\pi^4}{4} = 1.92 \; (2 \text{ marks for each term in the series}).$$

b) From the above, it can be seen that the derivatives follow the pattern: $-1, 0, 0, 1, -1, 0, 0...$

The general expression for the nth derivative is

$\dfrac{d^n x}{dt^n} = +1$ if n is a multiple of 4 (i.e., $n = 4m$, where m is an integer),

$\dfrac{d^n x}{dt^n} = 0$ if $n = 4m+2$ or $4m+3$, and

$\dfrac{d^n x}{dt^n} = +1$ if $n = 4m+1$.

c) We see from the above that the error term is $\pi^n/n!$

Since $\pi^2 \sim 10$, *we find that including terms up to the 5th derivative would reduce the error to less than* 1.

$\pi^8/8! \sim 10\,000/8! \sim 0.24.$

9.4 In the modified Euler method, $y(x_o + h) = y(x_o) + hf\left(x_o + \dfrac{h}{2}, y_o + \dfrac{hf(x_o, y_o)}{2}\right).$

A Taylor series expansion of the second term gives:

$$y(x_o + h) = y(x_o) + hf(x_o, y_o) + \frac{h^2}{2}\left[\frac{\partial f}{\partial x} + f(x_0, y_o)\frac{\partial f}{\partial y}\right] + ...$$

A Taylor series expansion of $y(x_0+h)$ gives

$$y(x_o + h) = y(x_o) + h\frac{dy}{dx}\bigg|_{x_o} + \frac{h^2}{2}\left[\frac{\partial f}{\partial x} + f(x_0, y_o)\frac{\partial f}{\partial y}\right] + ...$$

Chapter 10

10.1 If we replace the derivatives with the appropriate finite differences, we obtain:

$$2\left[\frac{u(x + h, y) + u(x - h, y) - 2u(x, y)}{h^2}\right] + \left[\frac{u(x, y + h) + u(x, y - h) - 2u(x, y)}{h^2}\right]$$
$$- \left[\frac{u(x + h, y) - u(x - h, y)}{2h}\right] = 2.$$

Rewriting this equation, we get:

$$u(x, y) = \frac{u(x + h, y)}{12}[4 - h] + \frac{u(x - h, y)}{12}[4 + h] + \frac{u(x, y + h) + u(x, y - h)}{6} - \frac{h^2}{3}.$$

Chapter 11

11.2. (a) If $r < 1$, the only fixed point is at the origin. This is because for small values of x, $r\tanh(x)$ increases similarly to rx, which will be below the $y = x$ line for $r < 1$.
   (b) A cobweb diagram indicates that it is stable (as seen below):

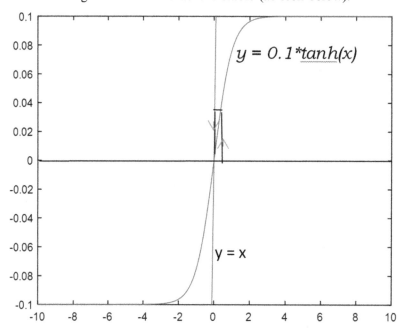

   (c) Since $r*\tanh(x)$ increases similarly to rx for small values of x, it is clear that there is only one fixed point for $r < 1$ and three fixed points for $r > 1$. Therefore,
$x*$
$r = 1$ is a bifurcation point.

The two new fixed points can be seen in the graph given below:

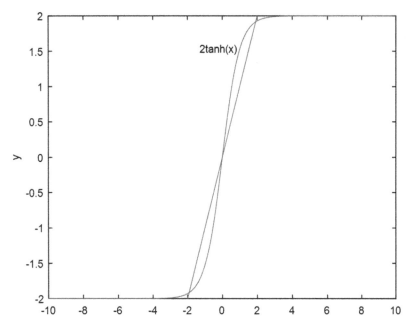

d) We can use a cobweb diagram to analyze the stability of the fixed points. Refer to the above figure. It can be seen that the two new fixed points are stable, while the fixed point at the origin is now unstable.

11.3 The graph for the sine map is similar to the graph for the logistic map. Both curves are smooth, concave down, and have a single maximum. Therefore, their bifurcation diagrams are similar.

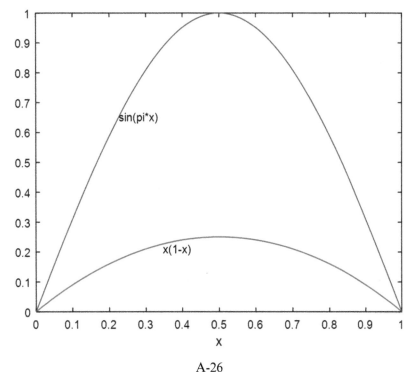

11.4 (a) To locate the fixed points, let us draw separate graphs of exp(x) and cos(x). The points where these functions have the same value are fixed points.

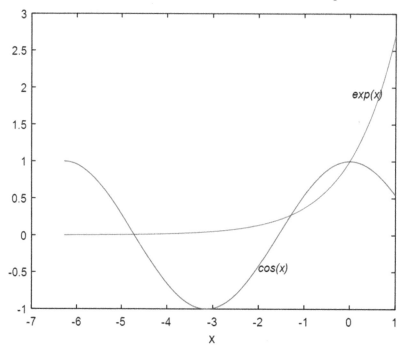

(b) We know that cos($-\pi/2$) = 0; therefore, exp($-\pi/2$)-cos($-\pi/2$) is positive, but exp(-1) $-$cos(-1) is negative. Therefore, an approximate value of the fixed point closest to the origin could be ($-\pi/2-1$)/2 ~ -1.28.

(c) $f'(0)$ = exp(0)+sin(0) = 1. Therefore, the fixed point at the origin is unstable.

(d) For positive values of x, exp(x) >1 and cos(x) < 1. Therefore, there are no fixed points for positive x.

11.5 (a) For $r > 1$, there is only one fixed point, which is at $x = 0$.

(b) As r decreases from ∞ to zero, the first value of r at which bifurcation occurs is $r = 1$. This is because for $r > 1$, there is only one fixed point, which is at the origin, but when $r = 1$, rx becomes tangential to sin(x) and when r is reduced to less than one, two new fixed points appear, as shown in the following graph.

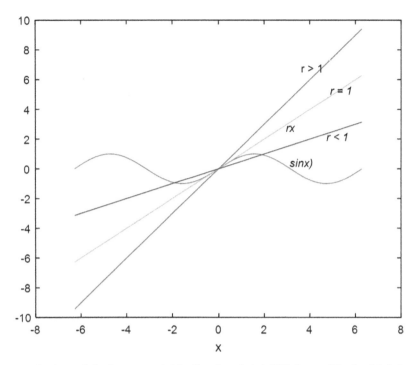

For $r > 1$, the origin is an unstable fixed point ($f'(0)$ is positive) which becomes stable when r is reduced to less than one. The two new fixed points that appear are both unstable.

c) The next bifurcation occurs when the straight line $y = rx$ becomes tangential to $y = \sin(x)$, as shown in the figure below:

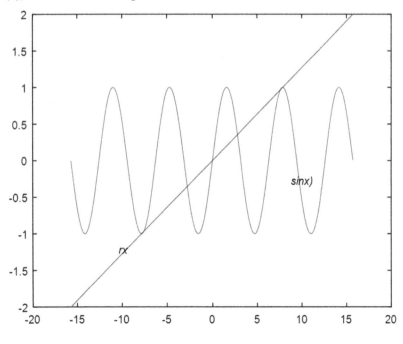

At this bifurcation point, $rx^* = \sin(x^*)$. In addition, since the straight line is tangential to the sine curve, $r = \cos(x^*)$. All the points at which both these equations are satisfied are bifurcation points.

11.6 The fixed point for this differential equation is at $x = 0$. To check the convergence to the fixed point, look at the derivative of this function:

$$f'(x) = \frac{1}{1 + x^2} - \frac{2x^2}{(1 + x^2)^2} = \frac{1 - x^2}{(1 + x^2)^2}.$$

The derivative is positive at the origin, becomes equal to 0 at $x = \pm 1$ and becomes negative beyond that.

Therefore, the Newton–Raphson method will only converge if the initial value of $|x|$ is less than one. An initial value greater than one would cause the subsequent values to move towards ∞, as seen in the figure given below:

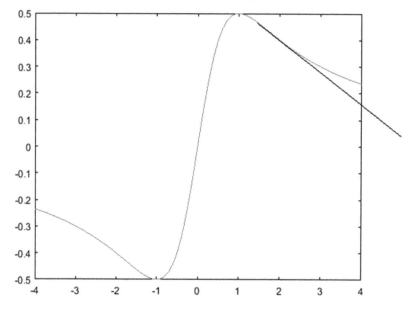

11.7 If only positive values of x (less than one) are chosen, then for $r < 4$, the values of x are always positive and in the range from zero to one.

If the sequence has to be random, then the logistic map should exhibit chaotic behavior. This happens when r is greater than 3.57. Of course, not all values of r greater than 3.57 lead to chaotic behavior, but most of them do.

11.8 The fixed points of this differential equation are those points where the first derivative is zero.

Therefore, $x^3 - 3x^2 + 2x = 0$
or $x(x^2 - 3x + 2) = 0$
or $x=0$, $x=2$, $x=1$.

To analyze the stability of these fixed points, we need to look at the value of the derivative of x^3-3x^2+2x with respect to x at the fixed point.

$f'(x) = 3x^2-6x+2$;

if $x = 0$, the derivative is positive, so the fixed point is unstable.

If $x = 1$, the derivative is negative, so this is a stable fixed point.

If $x = 2$, the derivative is positive, so it is an unstable fixed point.

11.9 The fixed points of this map are those points for which $x_{n+1} = x_n$. Therefore:

$$x_{n+1} = x_n = \frac{2x_n}{1 + x_n}$$

or $x_n(1+x_n)=2x_n$

$x_n = x_n^2$.

The fixed points are $x_n = 1$ and $x_n = 0$.

For stability, the derivative has to be less than one.

$$f'(x) = \frac{2}{1 + x_n} - \frac{2x_n}{(1 + x_n)^2}$$

For $x_n = 0$, the derivative is greater than one; therefore, this is an unstable fixed point.

For $x_n = 1$, the derivative is less than one; therefore, this is a stable fixed point.

CPSIA information can be obtained
at www.ICGtesting.com
Printed in the USA
BVHW012107120522
636361BV00003B/84